The Handbook of

SOFT FRUIT GROWING

OTHER GARDENING
BOOKS PUBLISHED BY
CROOM HELM

Growing Fuchsias
K. Jennings and V. Miller

Growing Hardy Perennials
Kenneth A. Beckett

Growing Dahlias
Philip Damp

Growing Irises
G.E. Cassidy and S. Linnegar

Growing Cyclamen
Gay Nightingale

Violets
Roy E. Coombs

Plant Hunting in Nepal
Roy Lancaster

Slipper Orchids
Robin Graham with
Ronald Roy

Growing Chrysanthemums
Harry Randall and Alan Wren

Waterlilies
Philip Swindells

Climbing Plants
Kenneth A. Beckett

1000 Decorative Plants
J.L. Krempin

Better Gardening
Robin Lane Fox

Country Enterprise
Jonathan and Heather Ffrench

The Rock Gardener's Handbook
Alan Titchmarsh

Growing Bulbs
Martyn Rix

Victorians and their Flowers
Nicolette Scourse

Growing Begonias
E. Catterall

Growing Roses
Michael Gibson

The Water Gardener's Handbook
Philip Swindells

The Salad Garden
Elisabeth Arter

The Pelargonium Family
William J. Webb

Hardy Geraniums
Peter F. Yeo

The Handbook of SOFT FRUIT GROWING

David Turner & Ken Muir

CROOM HELM
London · Sydney · Dover, New Hampshire

Croom Helm Ltd, Provident House, Burrell Row, Beckenham, Kent BR3 1AT
Croom Helm Australia Pty Ltd,
First Floor, 139 King Street, Sydney, NSW 2001, Australia

British Library Cataloguing in Publication Data

Turner, David
The handbook of soft fruit growing.
1. Fruit-culture – Great Britain
I. Title II. Muir, Ken
634'.7'0941 SB356

ISBN 0-7099-3538-2
ISBN 0-7099-2496-8 (pbk)

Croom Helm, 51 Washington Street, Dover, New Hampshire 03820, USA

Library of Congress Cataloging in Publication Data

Turner, David, 1918-
The handbook of soft fruit growing.

Includes index.
1. Berries. 2. Fruit—culture. I. Muir, Ken,
1925- II. Title.
SB381.T87 1985 634'.7 85-3812
ISBN 0-7099-3538-2
ISBN 0-7099-2496-8 (pbk.)

Typeset by Columns of Reading
Printed and bound in Great Britain by
Biddles Ltd, Guildford and King's Lynn

Contents

List of Figures

List of Tables

List of Colour Plates

Foreword

This book represents the pooling of the expertise of two professional growers, Ken Muir and David Turner. Their scientific approach to soft fruit growing, in particular the use of herbicides, irrigation and fertilisers, converted into practical terms is presented in a sound, down to earth manner. The normal range of soft fruits is fully described with the most up-to-date varieties mentioned and for the more adventurous gardener and those who like a challenge the unusual and newer kinds of fruits such as blueberries, cranberries and kiwifruit are included.

Ken Muir is a commercial fruit grower in his own right. He is also a very successful propagator and distributor of soft fruit plants to the amateur market. No other nurseryman in the last two decades has done more in introducing to the gardener the new varieties of soft fruits released from the British fruit research stations and from the Continent. In consequence the gardener has greatly benefited from the higher yield, enhanced fruit quality and increased resistance to pests and diseases which these new plants give. His forté, as is well known, is the strawberry and it can be said that his introductions to the United Kingdom have greatly enriched the strawberry industry.

David Turner, since his retirement from his post as Head of the Horticultural Department at Edinburgh University, has acted as a private horticultural consultant and established a soft-fruit nursery which specialises in the production of virus-free soft fruit stocks.

In these modern times of 'fast foods' reconstituted with flavour enhancers and added preservatives it is essential to have a balancing diet of fresh fruit to obtain the necessary vitamins. It could be argued that the easiest way to meet this requirement is to drive to the nearest 'pick-your-own' farm or visit the local greengrocer. This may be so but why do this if you have a garden? Why put money in their pockets? You can grow and pick your own. There is nothing more satisfying than growing your own, allowing it to ripen to its peak of perfection, using varieties with the flavours that *you* like, and experiencing that sense of achievement in having grown an economic crop. I am sure that this book will encourage and help the gardener to reach these admirable objectives.

H.A. Baker
The Fruit Officer,
RHS, Wisley

Preface

Methods of fruit production have been the subject of scientific investigation for many years, particularly at the East Malling and Long Ashton Research Stations and the Scottish Crop Research Institute. The object of these investigations was primarily to enable the commercial grower to produce better quality fruit at the lowest economic price for the consumer. Many of the new methods of production and control of pests and diseases apply equally well to the production of fruit crops in the private garden and allotment.

This book attempts to describe in simple language how these developments should be applied to the growing of soft fruits by the amateur gardener. Previously, fruit production was a very hazardous business or hobby. Plants were short lived, riven by pests and diseases so that yields were low and fruit quality poor.

The breeding of new varieties, the provision of healthy planting stock, new methods of cultivation and more effective pesticides have taken much of the risk out of soft fruit growing. As far as the authors are aware this is the first time that an attempt has been made to incorporate all these scientific developments in a book on soft fruit production for the non-professional gardener.

Acknowledgements

The authors wish to acknowledge the assistance of John King of the New Zealand Ministry of Agriculture in providing information about the cultivation of kiwifruit in that country.

They are also indebted to Murray Cormack for providing unpublished information about blueberry and cranberry trials carried out at the Scottish Crop Research Institute.

Colour Plates 4, 6, 7, 8, 9, 12 and 16 are reproduced by courtesy of the Director of the Scottish Crop Research Institute.

Chapter 1 General Considerations

Site

Cane and bush fruits can be grown over a wide range of soils and climatic conditions. Exceptions would be those sites that are directly exposed to salty winds coming off the sea, those at a height of 600 ft (180 m) or more and those where the soils are so shallow, infertile or overlie rock that local farmers do not attempt to cultivate them.

The choice of position in a garden is, in most instances, limited. Where there is a choice, the best position would be at the top of any slope, exposed to the full sun and not shaded by trees, hedges or walls.

Aspect of Rows

When rows of cane and bush fruits are sited so that they run from north to south, the fruit ripens evenly because both sides receive equal amounts of sunlight and warmth. If the rows are sited in an east to west direction, fruit situated on the north side of the bushes, and therefore shaded from the sun, will ripen two to three days later than that on the south side. These two factors, though not very important, should be taken into account when planning a garden. However, the slope of a garden and the shape of the fruit plot may be considered to be more important. Usually it is considered better to have the rows running up and down rather than across a slope. It is also more convenient for rows to be the same length as the breadth of the plot.

Shelter

The provision of shelter for fruit bushes is most important. This was demonstrated on one site which experienced medium exposure to the prevailing winds; over a seven year period strawberry yields were reduced on average by 32 per cent. Some of the beneficial effects provided by windbreaks are reduced water loss from the leaves and soil, a slightly higher temperature and humidity and a reduction in physical damage to leaves, flowers and fruit. Windbreaks also provide shelter for

Table 1.1: Suggested Living Windbreaks

		Planting Height		Planting Distance		E = Evergreen D = Deciduous	Notes
	Common Name	in	cm	ft in	m		
Alnus incana	Grey Alder	24-36	60-90	5 0	1.50	D	The roots fix nitrogen.
Berberis stenophylla		12	30	1 5	0.45	E	Can be allowed to flower before clipping.
Carpinus betulus	Hornbeam	24	60	1 5	0.45	D	Good for heavy soils. Trim during winter.
Cotoneaster simonsii		18	45	1.5	0.45	E	Semi-evergreen with red berries.
Cupressocyparis leylandii	Leyland's Cypress	12-24	30-60	3 3- 5 0	1.00- 1.50	E	Plant bare root plants and be prepared to stake.
Chamaecyparis alumii		12-24	30-60	3 3- 5 0	1.00- 1.50	E	Plant bare root plants and be prepared to stake. Bluish-grey and sea-green foliage.
Chamaecyparis fletcheri		12-24	30-60	3 3- 5 0	1.00- 1.50	E	Plant bare roots and be prepared to stake. Bluish-grey feathery foliage.
Fagus sylvatica	Green-leaved Beech	24	60	1 5	0.45	E/D	Retains dried leaves during the winter Do not plant on heavy wet soils.
	Purple-leaved Beech	24	60	1 5	0.45		
Hippophoe rhamnoides	Sea Buckthorn	18	45	2 5	0.45	E	Useful near the sea.
Ligustrium ovalifolium	Privet	24	60	1 0	0.30	E	Requires frequent cutting.
Populus taca-tricho	Poplar	48	120	4 0	1.20	D	Requires topping to restrict height.
Taxus baccata	Yew	12	30	2 0	0.60	E	Somewhat slow growing but makes a magnificent hedge.

pollinating insects leading to better setting of the fruit; unfortunately, these conditions also encourage pests and some diseases. If a living shelter is planted too near to the fruit bushes they will compete with each other for water and nutrients.

Shelter should filter the wind and reduce its speed; solid walls and wooden fences are unsatisfactory as they block the wind and this leads to violent turbulence on their leeward sides, which is more damaging than the original wind. Inert or living shelter that is 50 per cent permeable provides the most beneficial effect in reducing wind damage, an effect that will extend for a distance ten times the height of the windbreak.

If there is no shelter in a garden where a fruit plot is to be established this should be provided by erecting one of the proprietary propylene nets that are manufactured for this purpose; either on the side of the plot from which the prevailing winds blow or on all sides of the plot. At the same time a row of hedge plants should be planted on the leeward side of the artificial windbreak. By the time the hedge plants have grown to a satisfactory height the artificial windbreak will have deteriorated to such an extent that it should be removed.

In selecting a living windbreak it should be assumed that the spread of the roots will be equal to its height. Therefore, poplars, the fastest growing windbreak, should be planted only at the edge of a large garden when the fruit plot can be situated 30-40 ft (9-12 m) away. It can be expensive to top and prune poplars when they are fully grown. For a large garden, alders or *Cupressocyparis leylandii* are more suitable and sufficiently quick growing. There is no reason why the more attractive, more expensive but slower growing *Chamaecyparis allumii* or *fletcheri* should not be planted. For the smaller garden which is not likely to have a large fruit plot there is a wide variety of hedge plants that grow 6-12 ft (1.8-3.6 m) high. Table 1.1 gives details of various kinds of living windbreak materials.

The soil in which the windbreak is going to be planted should be prepared as carefully as that for the fruit bushes. Windbreaks should be kept weed free, given an annual application of compound fertiliser and watered when necessary if they are to provide shelter in the shortest possible time.

Planning

The number of bushes or canes of the various fruits that should be grown will obviously depend upon the size of the family household, their individual tastes and freezer capacity for the storage of surplus fruit for out-of-season consumption – not least the ability and inclination to undertake other methods of preserving. The possession of a deep freezer cabinet makes it unnecessary to have a hectic jam-making season when the fruit is ripe; instead, the frozen fruit can be made into jam at leisure and provide a superior product to one made in mid-summer and stored for several months, which inevitably loses some of its flavour and colour and can possibly become crystallised.

To assist in the planning of a fruit garden, the weights of the various fruits that a competent gardener, given reasonable soil, could expect to pick from fully-grown bushes, or per yd (m) of cane or vine, would be as follows:

	lbs per bush	kg per bush	lb per yd	kg per m
Blackberry	20	9.0		
Blackcurrant	8	3.5		
Blueberry	6	2.5		
Cranberry (per sq yd/m^2)			1	0.5
Gooseberry	10	4.5		
Hybrid berries	15	7.0		
Kiwifruit	65	30		
Raspberry (summer)	3	1.3	4.5	2.7
Raspberry (autumn)			1.5	0.9
Red and white currants	10	4.5		
Strawberry	1	0.5	2.0	1.1
Sunberry	16	7.0		
Tayberry	16	7.0		
Thornless Loganberry	12	5.0		
Tummelberry	16	7.0		

Bird Protection

The worst pests of soft fruits are marauding blackbirds and other species that eat the ripe fruits, and bullfinches, sparrows and other finches that strip the fruit buds from the branches during the winter. To prevent a proportion of the fruit from being lost each year, the only really effective remedy is to erect a fruit cage. This could be either a temporary tubular framework that is covered with a net during the fruit season, or a permanent structure that is covered with wire netting.

Rotation of Crops

One disadvantage of a permanent fruit cage is that it makes the rotation of crops difficult because the useful life of various crops will vary. However, this is not an insuperable difficulty. The first planting of strawberries could be followed, after 4-5 years, by a second planting on the same soil and, by the time the latter has reached the end of its useful life, i.e. after 8-10 years, one of the more permanent crops could be due for grubbing and could change places with the strawberries. Ten-twelve years later, it would probably be time for the blackcurrants and blackberries to be grubbed and change places. Alternatively after ten or twelve years a new fruit plot could be planted and a year later the fruit cage transferred to it so that there would be no break in the supply of fruit.

Preparation of Soil

Very few people are in the happy position to choose the house they live in because the soil on which it is situated is good for growing fruit. The majority have to manage the best they can with the soil they have inherited.

Soft fruits grow best on deep soils that provide a large volume of soil and sub-soil for the roots to exploit and obtain nutrients and water. The most common problem to be found in soils is a hard pan or a layer of soil that can no longer be penetrated by plant roots. Such pans could be caused by lorries running over the ground whilst the house was being built, repeated ploughing at the same depth when the ground was once farmland or by chemical reactions in the soil. A pan prevents surplus water draining away, resulting in the death of the roots. The soil below a pan will be a mottled red-grey or grey colour. A satisfactory test for determining the presence of a pan is to push a bamboo cane or thin metal rod down into the soil. The cane or rod should penetrate to a depth of 2-3 ft (60-90 cm) if a pan is not present.

Normally, a soil pan should be dealt with whilst preparing the plot for planting. Dig the top soil with a spade to a depth of 10-12 in (25-30 cm), and any hard pans in the 12 in (30 cm) below this should be broken with a fork. At the same time, particularly on light soils and gravels, the forking of a heavy application of bulky organic matter into the sub-soil will improve its fruit-growing quality. The organic matter should be mixed with the sub-soil and not placed in a thick layer at the bottom of the trench.

Heavy land that has a blue coloured sub-soil should be improved by the placing of a tile drain through the centre of the fruit plot at a depth of 2 ft (60 cm). The drain should be covered with 4 in (10 cm) of coarse gravel or cinders and empty into a ditch or soakaway (see Figure 1.1). A single drain will take surplus water away from a plot 33 ft (10 m) wide.

Perennial Weed Control

Many fruit species have the ability to continue cropping for 12-18 years, provided they are given satisfactory growing conditions; in

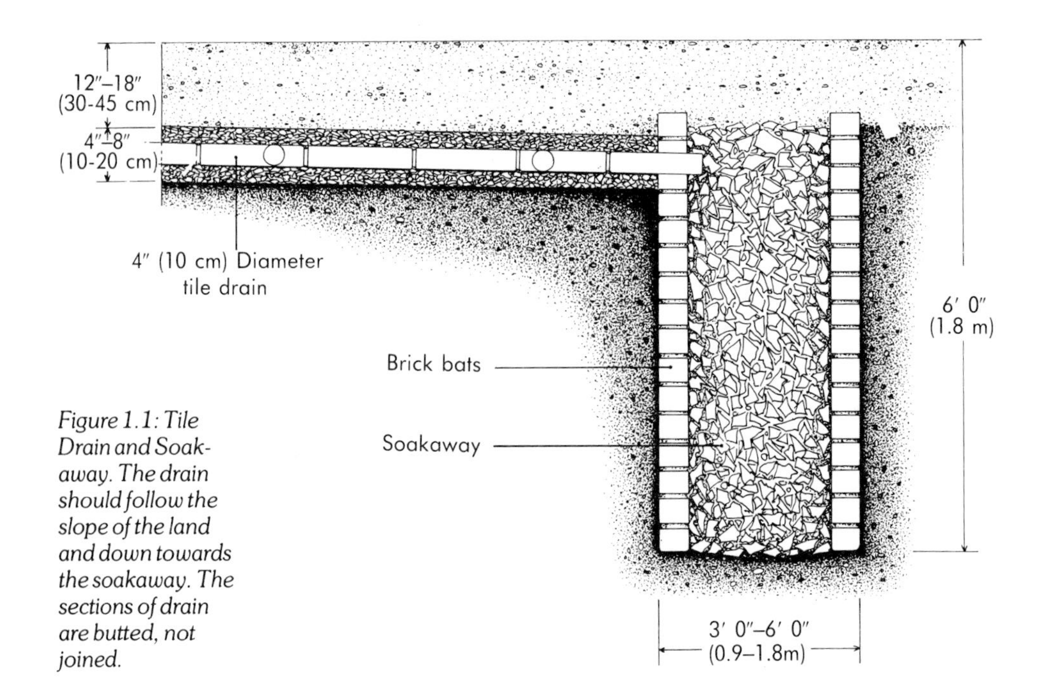

Figure 1.1: Tile Drain and Soakaway. The drain should follow the slope of the land and down towards the soakaway. The sections of drain are butted, not joined.

particular, freedom from perennial weeds. Whilst digging the soil in the manner recommended, the opportunity should be taken to make sure that the roots of any perennial weeds are removed and destroyed. In the United Kingdom, the perennial weeds that give the most trouble amongst fruit are common couch, convolvulus, broad-leafed dock, ground elder, horsetail and thistle. If the infestations are not too severe, it is possible to pick out every root whilst digging. If this cannot be done or the amount of root is too much to deal with in this way, the plot should be fallowed for one summer and treated with weedkiller.

The treatment consists, in February (in the UK), of making an application of one sachet of Weedex (simazine) in 2 gallons (9 litres) of water to 50 sq yds (40 m^2) with a watering can or spraying machine. This application will prevent the germination of annual weeds and allow the unrestricted growth of the perennial weeds. When these are fully grown, sometime in June, an application of 4 fl oz (113 ml) of Fison's Tumbleweed in ½ gallon (2.25 l) of water to 40 sq yds (34 m^2) should be made in the cool of the evening when there is a forecast of dry weather for at least 24 hours. Great care should be taken to ensure that none of the spray drifts onto cultivated plants outside the plot as it is likely to kill or severely impair those plants on which it settles afterwards. The watering can or spraying machine should be thoroughly washed out several times before being used for any other purpose.

Properly applied, Tumbleweed should kill over 95 per cent of the weeds. Within three to four weeks it should be possible to single out any weeds that are not going to die and they will require a second application. Finally, dig the plot over in early October (in the UK) so that the bushes may be planted when they are received from the nurseryman in early November onwards. Container-grown bushes and canes may be planted at any time during the year.

Liming

The acidity or alkalinity of a soil is expressed by the pH numbers 3-9 representing the hydrogen ions in the soil moisture. The lower the number the more acid and the higher the number, the more alkaline a soil. As the scale is logarithimic, a soil with a pH 5 is ten times more acid than a soil with a pH 6.

All fruit crops, with the exception of blueberries and cranberries, grow better and give the heaviest crops on soil that is slightly acid with pH between 6.0 and 6.5. At this level, all the major and minor plant nutrients will be more readily available to the plants; also, added fertiliser will be more effectively absorbed by the roots. Earthworms will thrive at this level assisting soil aeration and drainage. Where the soil has fallen to pH 5.5 growth of the bushes can be expected to be poor and below pH 5.0 they could fail altogether. The exceptions are blueberry and cranberry that require soil with pH 4.5.

The soil should be tested with a soil-testing kit or electronic meter and if found to be acid, the appropriate amount of lime that the chart indicates should be broadcast over the plot and rotavated or forked in. It takes twelve months for an application of lime to become fully effective.

Rates of application of carboniferous limestone, chalk or hydrated lime required to bring the pH to 6.5 in a medium loam soil are:

pH before liming	oz per sq yd	kg/m^2
6.5	nil	nil
6.0	7	0.25
5.5	14	0.50
5.0	26	0.90
4.5	37	1.25

A soil with pH 7.0 and above is alkaline and fruit crops growing in it may be expected to exhibit symptoms of manganese and iron deficiencies on their leaves, sometimes described as lime-induced chlorosis.

Manganese deficiency usually occurs on soils that have a high organic matter content, whilst iron deficiency occurs on soils that are low in organic matter. They may be distinguished by the fact that with manganese deficiency the chlorosis starts on the lower leaves whilst iron deficiency chlorosis starts on leaves at the tips of the shoots (see Plates 1 and 2). Temporarily the deficiencies may be remedied by the application of sequestered iron or manganese to the soil each February (UK).

The long-term solution to these troubles is to reduce the alkalinity of the soil, either slowly by applying nitrogen in the form of sulphate of ammonia or quickly by the application of flowers of sulphur.

The rates of application of flowers of sulphur required to reduce an alkaline soil to pH 6.5 are as follows:

	light sandy soil		sandy loam		heavy clay soil	
pH	oz per sq yd	g/m g/m^2	oz per sq yd	g/m^2	oz per sq yd	g/m^2
8.0 to 6.5	3	100	6	200	9	300
7.5 to 6.5	2	70	4	140	6	200
7.0 to 6.5	1	36	2	70	3	100

The sulphur should be mixed with the soil by forking or rotavating to a depth of 6-8 in (15-20 cm). If it is impossible to do this because too many roots would be destroyed, the application should be spread over three years. The soil should be tested one and two years later to check that the soil pH has been reduced to the correct level.

The various methods by which the soil can be acidified to make blueberry and cranberry cultivation possible are given on p. 53.

Nutrition

Soft fruits only require small amounts of nitrogen, phosphorus and potassium, and the soil requires only small amounts of lime. In spite of this, there is a mistaken tradition amongst gardeners, and by many horticultural writers, that large amounts of lime and fertiliser – in the form of bone meal, fish meal and inorganic fertiliser – should be automatically applied to soft fruit and other crops. Recommendations like 'apply three to four ounces per square yard of bone meal' are unnecessary and a waste of money. There is sufficient phosphorus in such an application to satisfy the requirements of any soft fruit crop for nearly 20 years.

The amounts of nutrients that are removed from the soil in the crop, leaves and prunings are relatively small. These should be replaced by comparatively modest applications of cheaper straight or compound artificial fertiliser. Many garden soils that have been under cultivation for a long period of time have become overloaded with potassium and phosphorus. Where this has occurred, the application of these nutrients should be reduced to a minimum or omitted altogether.

Organic Matter

There is little scientific evidence on which to base recommendations for the application of farmyard manure, compost or peat to fruit crops. The first two supply all the trace elements in addition to nitrogen, phosphorus and potassium, whilst peat contains negligible quantities. The chief values of organic matter are in increasing the water-holding capacity of soil in times of drought and improving soil structure so that it is easily workable and less inclined to restrict the draining away of excess water. It is suggested that the best policy should be, before planting, to incorporate one to two barrowloads of farmyard manure or compost to 6 sq yds (5 m^2) into the soil by rotavating or forking. If peat is applied a quarter of these amounts would be required. If the soil is not covered with a mulch of organic matter, a similar application should be applied in the sixth year.

Nitrogen

In general terms, the amount of nitrogen available to a plant controls the amount of vegetative growth it produces. When nitrogen is deficient, leaves tend to be small and pale green or yellow and shoot growth is restricted. Fruit yields are reduced as a result of the decrease in the number and size of fruits.

When excess nitrogen is given bushes are liable to be over-vegetative, with large, dark, blue-green leaves and the plants are more susceptible to infection. Fruit quality, especially taste and storage qualities, are also adversely affected.

During the first one or two years after planting the objective in applying nitrogen is to obtain rapid growth as quickly as possible and a large amount of wood that will later bear fruit. After this, just sufficient nitrogen should be applied to maintain the requisite growth of new shoots.

On cropping bushes it is recommended that too little nitrogen should be applied with a possible reduction in fruit size rather than too much that can lead to larger reduction in yield caused by disease infection and poor fruit bud quality.

Phosphorus

This element, like nitrogen, is concerned with the vital processes of plant growth and in particular with the formation of roots and the ripening of seeds and fruits. The amount of phosphorus taken up by fruit bushes is small and it is rare for the growth and yields to be adversely affected by a deficiency of phosphorus.

An application of 2 oz per sq yd (70 g/m^2) of superphosphate before planting and repeated every five years is all that is required for all the fruit crops.

Potassium

Potassium is very important in producing healthy, vigorous bushes. It exerts a balancing effect on growth and is said to increase resistance to diseases and to improve fruit quality. Its main effects are to assist chemical reactions and the translocation of food in the plant.

An acute deficiency of potassium in the soil will make itself evident as growth will be poor and a marginal scorching of the leaves will occur (see Plate 3).

On new gardens, broadcast 1 oz per sq yd (36 g/m^2) of sulphate of potash before planting, reduced to 0.3 oz per sq yd (12 g/m^2) for established gardens. This lower rate should be broadcast each March (UK) in subsequent years.

Magnesium

Magnesium is an essential constituent of chlorophyll, the green colouring matter in the leaves, concerned with the manufacture of carbohydrates. A deficiency of this element causes yellowing of the leaves. A deficiency of magnesium can be induced by excessive applications of potassium.

Many soils are naturally deficient in magnesium but crops may be kept adequately supplied by always applying lime in the form of ground magnesium limestone rather than the type of lime as recommended in the paragraph under lime.

Where a deficiency of magnesium does occur, this is best corrected by one application of 2 oz per sq yd (70 g/m^2) of Epsom salts (magnesium sulphate) in addition to normal applications of magnesium limestone.

Trace Elements

Trace element deficiencies occur very occasionally in soft fruit bushes in some areas. These deficiencies can be identified by visual diagnosis but should be checked before remedial action is taken. Toxicities of trace elements may arise as a result of soil acidity. Again an accurate diagnosis by an expert should be obtained.

When a specific trace element deficiency has been identified the recommended treatments are as follows:

Boron. Soil application of Borax at 0.75 oz per sq yd (27 g/m^2).

Copper. Foliar spray of 2 oz per gallon (12 g/l) of copper oxychloride in water plus wetting agent (if not already included with the copper oxychloride).

Iron. Foliar sprays of sequestrene (iron chelate) should be applied as soon as this deficiency has been accurately diagnosed.

Manganese. Two ounces per gallon (12g/l) of manganese sulphate in water, plus wetting agent, applied as a foliar spray. Two or three applications commencing in June (UK) may be needed each year.

Foliar Feeding

The application of cocktail mixtures of nutrients by spray on the leaves is not to be recommended. It has yet to be proven that foliar feeding serves any more useful purpose than soil applications, particularly as there is an insufficient quantity of any one element to correct a deficiency if one should be present. All the nitrogen, phosphorus, potassium and magnesium that crops require are better and more cheaply applied by a normal fertiliser programme.

Fertiliser Application

Different fruits require varying amounts of nitrogenous fertiliser depending upon their health, age, the length of the new growth they make each year and the lusciousness of their leaves. The success with which these requirements are gauged and satisfied governs the weight and quantity of the fruit that the bushes will bear. All these requirements can be met by applications of artificial fertiliser; there is no apparent advantage to be gained by applying the more expensive organic fertiliser. No matter in what form the manure is applied the roots take in nitrogen as nitrate or ammonium ions, phosphates as phosphate ions and potash as potassium ions, whatever the organic manuring school of thought may believe. Better results will be obtained by applying such straight fertilisers as nitro-chalk, superphosphate of lime and sulphate of potash. This enables the amount of nitrogen applied to be easily varied according to the growth of the bushes and the correct amounts of phosphatic and potassic fertiliser applied as recommended. For less effort it should be possible to obtain somewhat similar results, albeit at a higher cost, by applying one of the complete compound fertilisers such as National Growmore or Phostrogen (7% N_2 7% P_2O_5 20% K_2O).

A new garden that is being brought into cultivation for the first time should have broadcast over it after digging:

2 oz per sq yd (70 g/m^2) superphosphate
1 oz per sq yd (35 g/m^2) sulphate of potash

An established garden that has been under vegetable or fruit production for a number of years is very unlikely to require these additional applications. In both cases further specific recommendations for the application of fertiliser before planting and in subsequent years are made for each crop, in later chapters.

Method and Time of Application

In the United Kingdom, on newly planted crops, the fertiliser should be applied at the end of March or immediately after planting, whichever is the later, in a narrow band along the planting rows or in an 18 in (45 cm) circle round individual bushes. Top dressings of nitrogen should be applied in a similar manner in May or June and in some instances again in July. In established crops the fertiliser should be broadcast at the beginning of March 3 ft (1 m) either side of the rows of bushes or over the strawberry bed.

Mulching

The covering or mulching of the soil with dead organic materials such as straw, pulverised tree bark, peat or compost, prevents evaporation of moisture from the surface of the soil. A mulch will provide the equivalent of 2 in (50 mm) of rain in the summer. It will materially increase crop yields, particularly on lighter soils and in the drier eastern parts of the UK.

Some mulches will suppress annual weeds and may make the application of a herbicide unnecessary. However, it is considered that it is advisable, before putting a mulch on the soil, that a herbicide should be applied to make sure germination of annual weeds does not occur and grow through the mulch.

A mulch should be applied in late winter when the soil is

saturated with water. Straw should be applied to give a 4-6 in (100-150 mm) thickness, bark or peat a 2-3 in (25-75 mm) thickness and compost a 3-4 in (75-100 mm) thickness. Weed seeds will germinate on the surface of the compost. These materials gradually decompose making it necessary for small additional quantities of mulch to be added each year.

A mulch provides a clean surface on which to walk and increases the worm population but straw is a fire risk that could kill the bushes.

Another disadvantage of a permanent straw mulch is that it makes the flowers of fruit bushes more likely to be killed when spring radiation frosts occur in April and May. The reason for this is that when the air cools at night the bare soil radiates heat that it absorbed from the sun during the previous day. This heat can be sufficient to prevent the temperature falling to a level that kills the blossoms; 27°F (−2.8°C). The mulch acts as an insulating blanket and prevents the soil from giving up its warmth to air and the bushes.

The Purchase of Healthy Planting Stock

One of the main causes for soft fruit crops failing to make satisfactory growth and bear heavy crops is virus disease. These virus diseases differ from other kinds of plant diseases in two important ways. First, they are caused by agents too small to be seen even with the aid of an optical microscope: a gardener can only detect some viruses by means of the symptoms which they cause in plants. Secondly, once plants become infected by viruses, they remain infected and all such plants propagated vegetatively by suckers, runners and cuttings are themselves infected. Plants growing in the garden cannot be freed from virus infections, nor can they be protected from infection by means of chemicals, whilst routine spraying with insecticide is unable to prevent a fruit plot from becoming infected with virus, it does materially reduce the rate of spread within the fruit plot.

Virus diseases are spread from diseased to healthy plants by greenfly, leaf hoppers, mites and bees carrying infected pollen. The more important effects of virus infections are yellow streaking and spotting of the leaves, leaf distortion, reduction in vigour and loss of yield.

Unfortunately there are a number of viruses that do not always express their presence in the bush with symptoms on the leaves. These latent viruses affect varieties in different ways, for instance, raspberry leaf spot virus appears to have no effect upon the vigour of 'Malling Jewel' or give rise to symptoms on the leaves, yet when it is transmitted by aphids to 'Glen Clova' the leaves are distorted and covered with yellow spots; infected bushes die within three years.

The mosaic virus of raspberry, the symptoms of which are a violent yellow mottle along the lateral veins of the leaves, is caused by the combined infection of two viruses, rubus yellow net virus and black raspberry necrosis virus. Yet when either of these latent viruses infect a bush on their own, symptoms are not expressed by the leaves and their effect on growth and yield is negligible.

All fruit bushes may be affected by latent virus and the purchase of certified stock increases the useful life of bushes by reducing the risk of planting infected stock.

There are also a number of so-called soil-borne virus diseases; these, in fact, infect weed seeds and the weeds that grow from them. Infected weeds bear infected seeds that will maintain the infection in a plot of ground. The viruses are transmitted by eelworms (small transluscent, threadlike creatures) from infected weeds to other weeds and also to crop plants. Infected fruit bushes may exhibit symptoms on their leaves that are not dissimilar to the virus diseases spread by greenfly; the bushes may be killed by the virus or the bushes may be tolerant to, and show no signs of, the infection. The one distinguishing feature of soil-borne virus infection of a susceptible variety is that the infected bushes occur in small patches that gradually spread outwards.

The only way to prevent the worst effects of virus infections is to purchase healthy plants in the first place in the knowledge that experience has shown that they will crop satisfactorily for a number of years before new virus infections have too serious an effect on growth and yield. For this reason, it is important that plants should be purchased from nurserymen who sell stocks that have been certified by the Ministry of Agriculture or the Department of Agriculture for Scotland as being healthy and true to type.

There are three types of certificate with which bushes are sold:

Special Stock Certificate

Bushes with this certificate have been propagated from plants that were virus free when they were obtained from one of the Research Stations or the National Seed Development Organisation. They would be the healthiest stock available and are primarily sold for further propagation or for the establishment of commercial fruiting plantations. Nurserymen sometimes have bushes with this certificate for sale to their customers.

'A' Certificate

The majority of certified bushes for sale have this grade of certificate. Such bushes would have been propagated from plants that on planting possessed a Special Stock certificate. They should be perfectly satisfactory for planting in gardens and every effort should be made to purchase planting stock with this grade of certificate. It should be possible to purchase blackcurrant, the gooseberry varieties 'Invicta' and 'Jubilee', Loganberry, raspberry, the redcurrant variety 'Redstart', strawberry, Sunberry, Tayberry and Tummelberry planting stock with this standard of certificate, now or in the near future.

'H' Certificate

The 'H' certificate was instituted in 1983 to indicate a satisfactory standard of health of certain stocks of strawberry, blackcurrant, gooseberry and redcurrant, that are not eligible for an 'A' certificate. This enables a certificate for health to be issued for varieties that have been bred in foreign countries and are being grown in the United Kingdom for the first time; also for varieties that have been micro-propagated. The certificate is for apparent freedom from virus and other diseases but does not guarantee that a variety is true to name. Bushes with an 'H' certificate should only be purchased when those with an 'A' certificate are unavailable.

Some fruit varieties are offered for sale with the prefix 'Medana', both in the UK and the USA, and also in the EEC. This is a registered trademark that indicates the variety has incurred a royalty payment and that it was raised at a government-financed Research Institute.

Do not accept gifts of strawberry runners, raspberry canes, blackcurrant bushes or cuttings that have been propagated by friends, as they are most likely to be infected with virus diseases. Diseases such as red core on strawberries, big bud mite on blackcurrants and crown gall on raspberries and blackberries may also be infecting such gifts. Once a soil has become infected with red core the disease cannot be eradicated.

Plant Quality

British Standards are laid down to describe the health, size and physical attributes of good quality plants. It is considered these are of limited value in giving descriptions of good root systems and that sizes and numbers of the branches can only be described as minimal. The number of nurserymen who sell bushes with the BS standards label is limited, therefore the majority of gardeners who wish to ensure that they are supplied with first quality plants should, in the United Kingdom, order from a nurseryman recommended by *Which?*; from one who guarantees to replace any plants that die; or one that experience has shown always sells good quality stocks.

Excerpts from British Standards for Soft Fruit Plants. These excerpts from B.S.3936: Part 3 are reproduced by permission of the British Standards Institution, 2 Park Street, London W1 from whom complete copies of the document can be obtained.

(9) The nursery stock shall be substantially free from pest and disease and materially undamaged. No roots shall be subjected, between lifting and delivery, to adverse conditions such as prolonged exposure to drying winds and frost.

(10) Packaging shall be adequate to protect the plants and prevent them from drying out.

(11) The plants shall comply with the appropriate requirements laid down in Section II for certification and plant health.

(14) Each plant, bundle or batch of plants shall be labelled with the name and variety of fruit; the name of the supplier unless otherwise agreed.

(17) Blackcurrants – If a variety is included in a certification scheme operated by one of the U.K. Departments of Agriculture, the plants shall be certified stock. One-year-old plants shall have been propagated from certified stock.

(17.2) Blackcurrant bushes shall have all their branches coming from below or near ground level. They shall be visibly free from buds affected by Big Bud Mite. One-year-old plants shall have at least two vigorous shoots, each not less than twelve inches (30 cm) in length. Two-year-old or three-year-old bushes shall have not less than three shoots, each not less than 16 inches (40 cm) in length.

(18) Redcurrant and white currant bushes shall be on a leg six inches (15 cm) to ten inches (25 cm) long and two-year-old bushes shall have not less than three branches.

(19) Gooseberry bushes shall be on a leg six inches (15 cm) to ten inches (25 cm) long, unless otherwise stated by the supplier, and two-year-old bushes shall have not less than three branches.

(20) Raspberry canes shall have been certified under a raspberry certification scheme operated by one of the U.K. Departments of Agriculture, or shall have been derived from canes so certified within the previous twelve months.

(20.2) Raspberry canes shall be well rooted and shall be not

Figure 1.2: A Well-grown One-year-old Hybrid Cane Fruit, Before Planting

less than 25 inches (62.5 cm) in length as grown.
NOTE. The canes may be shortened for despatch.

(21) Blackberry and loganberry canes shall have well developed fibrous roots and well developed basal buds.

(22) Open-ground strawberry runners shall have been certified under a strawberry certification scheme operated by one of the U.K. Departments of Agriculture. Runners that have been rooted by mist propagation shall have been obtained direct from a strawberry runner bed similarly certified. Plants despatched in August and September may be sold as 'Entered for certification' if the final inspection has not been completed at the time of sale.

(22.2) Strawberry plants shall be runners taken from parent plants which shall be not more than one year old. Open-ground runners shall have a good crown and a well developed root system. Mist propagated plants shall have roots sufficiently developed to hold their ball of soil, and shall have a good crown. They shall have been grown in pots of sufficient size to maintain the plants in normal growth.

Importance of Correct Handling of Plants

When plants are received, the ideal would be to plant them immediately, if the soil has been prepared and it is in a plantable condition: that is, dry enough to be firmed without turning into a sticky mess. Otherwise, heel the plants in a trench, having untied the bundles, spaced out the roots and lightly covered them with damp soil. If the ground is frozen, open the package and make sure that the roots have not dried out, moisten them if necessary and then re-close the package and store it in a cold shed until the soil thaws out.

Time of Planting

In the United Kingdom, fruit bushes, but not strawberries, may be planted at any time during the winter whilst they are dormant but growth in the following summer will be better if they are planted before Christmas. Bushes should be ordered early in the autumn from a nurseryman with the stipulation that if possible, they should be delivered before the end of December. They should be planted as soon as the soil is dry enough and can be worked into a rough tilth so that the soil can be firmed over the roots.

Strawberry planting may commence in March (in the UK) as soon as the soil is dry enough to work into planting condition; planting may continue until mid-October in the north and mid-November in the south of the country (UK). They should not be planted during the winter as frost will lift them out of the ground. Nevertheless, delivery of plants may be taken during the winter provided they are potted up or heeled in a cold frame, for planting in the spring.

Fruit Plot Management

Preparation for planting should take place in early autumn when the soil should be dug to a spade's depth and the bottom of the trench forked up. This will be the last time for several years that it will be possible to work the soil to this depth and it should be done thoroughly and not skimped. Rotavators are neither powerful enough, nor designed to cultivate the soil down to this depth.

This early preparation allows the soil to settle and form a natural tilth before planting commences in November or December (UK). Immediately prior to planting, when the soil is sufficiently dry, the final rough tilth should be prepared by raking, forking or rotavating. Experiments at Research Stations in the United Kingdom have shown conclusively that all kinds of soft fruits, once planted, grow more satisfactorily and produce the heaviest crops when grown in soil that is not disturbed by further cultivation, once the initial soil preparation has been undertaken. The reasons for this are that the top soil contains the highest concentration of plant nutrients and is more retentive of soil moisture. Cultivation to destroy weeds also destroys the crop roots that would otherwise be taking in water and nutrients.

Even weeds that are allowed to grow for a short time seriously compete with the crop and reduce growth and yield. Weeds are particularly harmful when they compete during April, May and June with newly planted bushes.

There are two ways of avoiding root destruction. The first is to repeatedly hoe, very shallowly, with a dutch hoe; this may be feasible in theory but it is difficult in practice. A rotavator, where it is able to work between the bushes, can be used for this provided that it is operated on every occasion at the same shallow depth.

The second and easier way is by the application of herbicides.

Weed Control

Fortunately there are herbicides available that enable weeds to be controlled satisfactorily in all the soft fruit crops. Gardeners who are prepared to carry out the application instructions carefully should always use these herbicides.

Weedex (simazine). Weedex is applied to and absorbed by the top inch (2.5 cm) of soil. The roots of germinating weed seedlings absorb Weedex and are killed by it. To be effective against annual weeds and not harm the bushes, the following rules should be observed.

(1) For winter application the best time in the United Kingdom is February in the south and March in the north of the country when the soil is moist and rainfall can be guaranteed to wash the herbicide into the soil before weeds begin to germinate.

(2) The soil should be moist, have a firm, fine tilth and should not be cloddy.

(3) The soil should be free from all weeds. If weeds are present, Weedol (paraquat) should be applied at the same time, provided the restrictions on its use with blackcurrants and gooseberries are observed.

(4) The annual weeds – cleavers (*Gallium apurine*), knotgrass (*Polygonum aviculare*), buttercup (*Ranunculus repens*), redshank (*Polygonum persicaria*) and groundsel (*senecio vulgaris*) – may be resistant to Weedex, so, in these circumstances, they should be killed by Dutch hoeing and not allowed to seed.

(5) If the soil has not been completely cleared of perennial weeds before planting they will over-run and smother the whole plot. However, they can be dealt with by repeated hoeing until the food reserves in their roots have been exhausted or by painting their leaves with a brush dipped in a mixture of nine parts of wallpaper paste and one part of Fison's Tumbleweed. This herbicide is lethal on fruit bushes so should not be allowed to come in contact with their foliage.

(6) Under normal circumstances in the autumn following an application of Weedex, moss will start colonising the soil surface. This moss should be allowed to flourish and form a complete carpet; it is not harmful to the bushes and makes an excellent surface on which to walk.

(7) Weedex should only be applied to established strawberries but they are more easily damaged by incorrect application than the other fruits.

Casoron G (dichlobenil). This herbicide is available as small granules and is intended primarily for the control of certain perennial weeds in some of the soft fruit crops. It will also prevent the germination of many annual weeds. Casoron G should not be used in strawberries and only applied to the other soft fruits when they are two years or older. The best time of application is January, well before the plants show any signs of growth. The herbicide may be used for spot application to individual weeds during the spring and summer but this is not so effective compared with a winter application.

Broadcast at 1/6 oz sq yd (6g/m^2), the majority of annual weeds will be prevented from germinating and moderate infestations of established annual weeds will be killed. Applied at 1/3oz per sq yd (12g/m^2) Casoron G severely checks or kills colt's-foot, couch grass, dandelion, dock, ground elder, horsetail, thistle and willowherb.

One year should elapse between the application of Weedex or Casaron G before the planting or sowing of seeds of another crop. When digging the soil in preparation for such a crop, the soil should be inverted as far as this is possible. Ideally, the top 2 in (5 cm) should be removed first and put in the bottom of the trench before digging each strip of soil.

Covershield (propachlor). This is another granular herbicide that can be used with all fruit crops but as it is only effective for eight weeks its use should be restricted to strawberries, as Weedex is a better herbicide for the other crops. Covershield should be sprinkled on the soil immediately after the strawberries have been planted at the rate of 1/8 oz per sq yd (4.5 g/m^2). Further applications should be made at eight-weekly intervals to keep the strawberries free from weeds. Covershield will not control common fumitory (*Fumaria officinalis*), charlock (*Sinapsis arvensis*) and field pennycress (*Thlaspi arvensis*).

Similar applications can be made to fruiting strawberries each year from March onwards.

Weedol (paraquat). Weedol is a contact herbicide that dessiccates the green parts of any plant with which it comes into contact: it is inactivated on contact with the soil, leaving no residues to affect the roots or future crops. The rate of dilution is one sachet in 1 gallon (4.5 l) applied to 20 sq yds (17 m^2) by means of a sprinkler bar attached to a watering can or with a spraying machine, provided the latter can be adjusted to give a coarse spray that will not drift on to the crop plants.

Weedol should be applied amongst cane fruit bushes during the dormant period, to gooseberries before the middle of January (in the UK) when the buds begin to move, and only as a directed spray, and under bushes of blackcurrant and redcurrant or the buds will be killed. It can be used throughout the summer for the spot treatment of weeds, provided none is allowed to fall on the leaves of the plants.

Weedol should be used in strawberries after harvest, to kill surplus runners growing between the rows, provided great care is taken to ensure that none falls on the rows. Two applications are needed, one in early September and the second in late October. The stolons of the runners should be cut with a spade or turf edging iron before making an application, to prevent translocation of the herbicide from the treated runners to the parent plants.

Watering

In the eastern counties of England drought conditions occur seven years out of ten; in Scotland and the wetter parts of the United Kingdom, only three years out of ten. In these dry summers the watering of fruit bushes increases yields and improves growth, the latter ensuring that the crop in the following year will be satisfactory. The shallow light soils require watering more frequently with larger quantities than the deeper heavier soils.

The cheapest satisfactory way of applying water is with an oscilating garden rainer that applies water to a square or rectangular area of soil. Rotary sprinklers are less satisfactory because of difficulty in obtaining the correct overlap of adjoining patterns. Application of water with a hosepipe or watering can is impracticable because of the large quantities of water required, the difficulty of getting an even application and because of run-off.

The times during which water should be applied to the different fruit crops is fairly critical. If water is applied unnecessarily or at other times, the yield of fruit can be reduced, growth can be adversely affected and infections by disease encouraged. On very light sandy and gravelly soils water may be applied outside these times when the leaves are wilting.

It is possible to calculate fairly accurately how much water to apply during periods of drought but the necessary information is not available to amateur fruitgrowers. A rule of thumb is, on light soils of the 18-24 in (45-60 cm) depth, to apply 4½ gallons of water per sq yd (25 l/m^2) at the end of 14 days of drought and make a similar application every seven days for the period that the drought continues. On deeper heavier soils, delay watering until the third week of the drought when 9 gallons of water should be applied to each sq yd (50 l/m^2) and whilst the drought continues, apply a similar quantity every 14 days.

Four and a half gallons of water per sq yd (25 l/m^2) is the equivalent of 1 in (2.5 cm) of rain. A reasonably accurate measurement can be made by placing two or three open straight-sided tins under the oscilator or sprinkler and measuring the depth of water with a ruler.

The periods during which water should be applied to the various crops are given in the chapters detailing the cultivation of each crop.

A plastic rain gauge can be purchased cheaply from horticultural shops. A record of rainfall should be kept and this, combined with the knowledge that fully grown bushes when in leaf take 4½ gallons of water from each sq yd (25 l/m^2) of soil or 1 in (2.5 cm) each week, can give a useful indication of when and how much water to apply.

Propagation

With the exception of gooseberries, all the soft fruits are fairly easy to propagate, so that many gardeners are tempted to multiply their own stocks either for planting themselves or for giving to their friends. Unfortunately, fruit bushes grown in urban gardens become quite rapidly infected with virus diseases carried by aphids, and blackcurrants with reversion virus carried by big bud mite. Strawberries and raspberries can become infected with soil-borne virus transmitted to the roots by free-living eelworms, and strawberry plants can have their roots and crowns infected by red core (*Phytophthora fragariae*), verticillium wilt (*Verticillium dahliae*) and crown rot (*Phytophthora cactorum*) diseases. This is not a complete list of troubles with which propagating material can become infected and from which it cannot be freed.

Although it is strongly recommended that gardeners should not accept plants (with the possible exception of redcurrants and gooseberries) as gifts that have been propagated by their friends, the methods by which these fruits are propagated will be outlined in later chapters.

Pests and Diseases

The pests and diseases that regularly or very occasionally attack soft fruits are described in two later chapters, together with the control measures which should be taken to reduce to a minimal level the damage they cause. The descriptions given of the organisms, and the damage they cause, should enable a correct identification to take place.

Gardeners often think it is only other people's fruit that is attacked by pests and diseases and fail to recognise the troubles that are affecting their own crops. Even when the realisation dawns that they have a problem on one of their own crops, it is surprising how often the wrong chemical is used and as often as not at the wrong concentration and at the incorrect time. Gardeners can, on occasions, be misled by advice given in ironmongers, garden shops and garden centres; the most reliable sources of advice are The Royal Horticultural Society, Wisley, Surrey, Colleges of Agriculture and Horticulture, horticultural consultants, lecturers at evening classes on gardening and the manufacturers of horticultural chemicals. (See pp. 172-7).

Having made sure that the trouble has been correctly identified, the correct chemical for its control should be obtained together with the written instructions that should state unequivocally that it does control the particular pest or disease. The instructions regarding dilution and time of application must be followed.

Spraying

The objective of spraying is to distribute a very small amount of chemical evenly over the stems, leaves, flowers or fruits. The most effective way of doing this is to mix the chemical with water and apply the mixture with a spraying machine so that the entire surface of the bush is wetted until drops of liquid begin to run off the leaves. Spray when there is little wind, using a fine spray so that none of the spray is blown away onto other crop. A wetting agent is always added to chemicals to enable the spray to wet the leaves, etc. If plants are very hairy or have a waxy bloom there could be insufficient wetting agent and the liquid simply runs off the leaves. If this is seen to be occurring, add ⅕ fl oz (5 cc) of washing-up liquid to each gallon (4.5 l) of spray.

A spraying machine suitable for a small fruit plot would hold about 1 gallon (4.5 l). One for a larger garden would hold between 2-4 gallons (9-18 l) and be carried on the back, and is effective for applying pesticides to bushes or herbicides to the soil.

Insecticidal and fungicidal dusts are also available but they are most expensive and less effective than sprays. They should be applied early in the morning or late at night when there can be little wind and bushes are likely to be covered with dew. They are usually sold in an applictor and are very convenient for dealing with small outbreaks of pests or diseases when there may not be time or the necessity to fill a spraying machine.

Chemicals that are available can be poisonous but no more so than the nicotine, lead arsenate and mercurous chloride that were in general use before modern compounds were developed. All chemicals should be kept under lock and key and well out of the reach of children. They are expensive so should not be wasted by mixing them incorrectly with water. Too high a concentration is not only wasteful but could have a damaging effect when applied to the bushes. Too low a concentration will result in inadequate control of the pest or disease. Although some pesticides are supplied in measured quantities or the container cap can be used as a measure, a measuring cylinder and a small weighing machine should be part of a gardener's equipment. The instructions supplied with the chemical should always be read and followed in detail. In particular, the time that should elapse between spraying and harvesting of the crop should invariably be observed, otherwise a chemical deposit could be present on the fruit when it is eaten.

Chapter 2
Blackberries and Hybrid Berries

Introduction

The weight of fruit that these kinds of bushes bear depends mainly upon the length of season over which they ripen their berries. This is governed largely by when the first autumn frosts occur and prevent the ripening of any further green berries. Although some late-ripening varieties are potentially high yielding, it is the early-ripening varieties that should be relied upon to bear heavy crops every year. The further north and north east a garden is situated in the United Kingdom, the less worthwhile it will be to grow late-ripening varieties. Conversely, the heaviest yields are produced in the south of the country. These fruits grow best in zones of hardiness 5-8.

The least space that any of these fruits occupy is 8 ft (2.5 m) between bushes in the row. Where space is limited it is suggested that, for blackberries, 'Ashton Cross' should be grown in the south and 'Bedford Giant' in the north. Of the hybrid berries, the choice should be between Tayberry and Thornless Loganberry (LY 654), the latter only being chosen if it is essential to grow a bush with thornless canes. Where space is available, it is suggested that thorned or thornless Boysenberry should be grown for its high yield of large characteristically-flavoured berries. 'Oregon Thornless', a late-ripening blackberry, has deeply segmented 'parsley' type leaves that exhibit attractive coloured tints in autumn. This variety should also be planted whenever possible.

Varieties – Blackberries

'Ashton Cross'

This mid-season variety is a true blackberry, selected by the Long Ashton Research Station (LARS) from wild blackberries growing in the woods in the West Country. It is extremely heavy yielding, bearing round, moderate-sized berries. Though somewhat acid, they have the typical flavour of wild blackberry. The plants are vigorous with thin, wiry thorny canes.

'Bedford Giant'

Although 'Bedford Giant' is reputed to be a true blackberry, it may well be a hybrid raised by Laxton of Bedford using Veitchberry as the female parent. It is the earliest blackberry to ripen, bearing very large bright black berries with a slightly acid but weak blackberry flavour. The canes are vigorous, thick and thorny.

'Himalaya Giant'

This mid-season variety can usually be relied on to crop heavily. The berries are large jet black in colour, acid and of moderate flavour. They are borne in large numbers on long laterals. The canes are massive and vigorous with large sharp, thorns. This is not a very suitable variety for small gardens.

'Oregon Thornless' (Parsley or Cut Leaf)

'Oregon' ripens two weeks after 'Himalaya' so the yield of fruit will be the lowest and most variable of the blackberries. The fruit has a good blackberry flavour, is of medium size with large shining drupelets. The canes are moderately vigorous.

Varieties – Hybrid Berries

Boysenberry

The origin of this hybrid is not known with any certainty. It was selected in 1920 and became generally grown after 1935. Boysenberries are grown extensively in New Zealand and the USA where heavy crops are produced. It grows well and is hardy in the south of the United Kingdom. The berries are large, round-oblong in shape and purplish-black in colour when ripe. They have a characteristic attractive flavour. Thorny and thornless varieties are available.

Loganberry

The Loganberry was derived from an accidental crossing of the raspberry 'Red Antwerp' and a blackberry in California in 1880. A thornfree chimaera was discovered in 1929 and, apart from the absence of thorns, is identical with the original thorny strain. Virus-free stocks of thorny and thornless Loganberries were introduced by East Malling Research Station (EMRS) and are now referred to under the codes LY 59 and LY 654 respectively. In the United Kingdom, Loganberries ripen towards the end of the first week of July and some berries will still be ripening towards the end of August. The berries are large, blunt-conical in shape, dull dark red in colour and very acid.

Sunberry (EMRS)

The Sunberry is a cross between the raspberry 'Malling Jewel' and *Rubus ursinus*, and having completed its trials satisfactorily is just becoming available. The berries are similar in size to Loganberry, conical in shape and, when ripe, very dark purple-black in colour. The flavour is good. It ripens a few days after, and finishes ripening before, Tayberry. The growth is vigorous with strong spiny canes that makes training and picking difficult which makes the alternate bay system of training advisable (see page 32). It is not a variety to plant in a small garden.

Tayberry (SCRI)

The Tayberry is a hybrid arising from crossing the American blackberry variety 'Aurora' and an unnamed Scottish Crop Research Institute raspberry seedling. Plants have been on sale for five years and Tayberry is proving to be a popular new fruit. In the UK, the berries commence ripening during the first days of July and continue doing so until mid-August. They are much larger than those of Loganberry, blunt chisel-shaped and dark red in colour when ripe. They are less acid and more highly flavoured than Loganberry, having inherited the rich blackberry flavour of its American parent. The hard plug disappears when the fruit is frozen, bottled or made into jam; all these products have a bright attractive appearance and an excellent flavour. The new canes are only moderately vigorous and prickly. The fruit is borne on short laterals and easily picked. (See Plates 4 and 5).

Tummelberry (SCRI)

The Tummelberry is an even newer hybrid fruit obtained by crossing Tayberry with one of its unnamed sister hybrids. Compared with Tayberry, the fruit is a brighter red colour and a sharper and less aromatic flavour. The shape is less angular and it resembles a very large raspberry, though the plug remains inside the berry. The season of ripening is a week later than that of Tayberry. The canes are slightly more erect and hardy than those of Tayberry, otherwise it gives a similar or slightly lower yield. Plants should be available for purchase after 1985. (See Plate 6.)

Planting Stock

'Ashton Cross', 'Bedford Giant', Boysenberry, Sunberry, Tayberry and Tummelberry should become available as Special stock or 'A' Certificate stock in the immediate future.

The 'Medana' suffix will be seen on the labels of some stocks of plants of 'Ashton Cross', Sunberry, Tayberry and Tummelberry. 'Medana' indicates the variety was raised at a British Research Station and that the plants being purchased originated from virus indexed stock and were certified as healthy by the Ministry of Agriculture during the preceding summer.

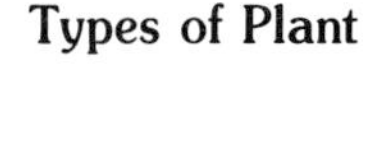

Types of Plant

All varieties of black and hybrid berry should be available during the winter as one-year-old plants with bare roots. These have a single stem that branches into two or more strong canes which will have been cut back by the nurseryman in order to facilitate easy handling. They should have a large fibrous root system (see Figure 2.1).

Depending upon the method of propagation, Tayberries, may also be offered as canes that resemble good quality raspberry canes.

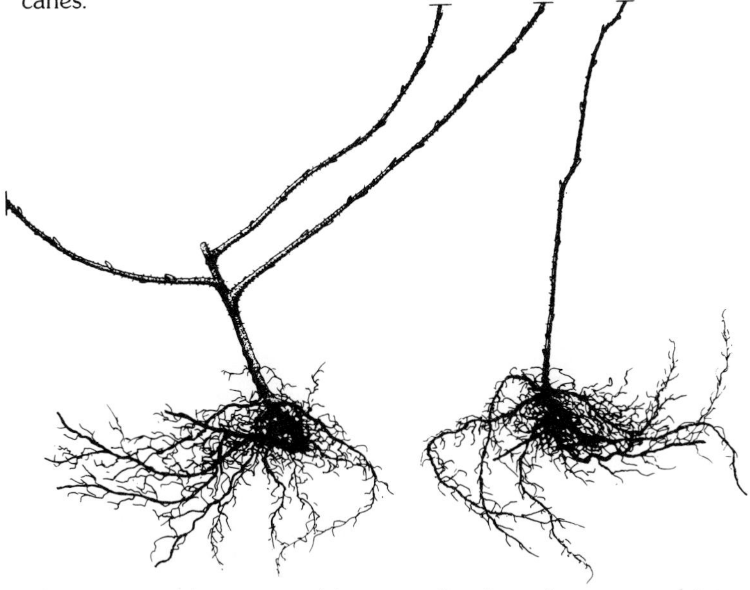

Figure 2.1: Examples of Good One-year-old Tayberry Plants

Sometimes, plants are sold as tips that have been rooted into 3½ in (9 cm) pots. Although small, these are good value for money but take a year longer to bear a full crop.

These berried fruits can also be purchased from garden centres during the summer as containerised plants in full growth. They could vary in size from plants with a single cane a few inches (cms) high to large plants with six canes several feet (1-2 m) tall.

Planting Distances

The distances that plants should be spaced in the row on average soils are as follows:

BLACKBERRIES AND HYBRID BERRIES

	ft	m
'Ashton Cross'	12	3.6
'Bedford Giant'	12	3.6
'Himalaya Giant'	15	4.5
'Oregon Thornless'	6	1.8
Sunberry	15	4.5
Tayberry	8	2.4
Thornless Boysenberry	8	2.4
Thornless Loganberry	8	2.4
Tummelberry	8	2.4

If more than one row is planted, for the weaker varieties such as Tayberry, 'Oregon Thornless' and Boysenberry, the distance between the rows should be at least 6 ft (1.8 m) and preferably 7 ft (2.1 m), whilst 'Bedford Giant', 'Himalaya Giant' and Sunberry would not be too far apart at 8 ft (2.4 m) to accommodate the long laterals and leave room for walking between them. If a fruit cage is too small to accommodate these row widths, narrower spacings could be adopted but there would be some difficulty experienced in tending the plants and picking the fruits.

Planting

Where possible, bare root plants should be purchased and planted during November and December (in the UK), though careful planting during the late winter can be satisfactory. Plants in containers may be planted at any time of the year. Those that are planted whilst in leaf should be watered before planting and afterwards watered with a garden sprinkler until their roots have established themselves in the soil and the canes have continued to grow. A hole should be dug sufficiently large either so that the roots of plants can be spread out or to take the root ball. The depth should be such that white growth buds should be not more than ½ in (1.75 cm) below soil level and the surface of the root ball should be at soil level. The soil should be firmed on the roots or round the root ball with the sole of the boot. All dormant plants planted during the winter should have their canes cut back to 12 in (30 cm) in length before planting. Plants purchased in containers should be full of roots and left unpruned and allowed to bear fruit.

Manuring

These fruits require generous applications of fertiliser if they are to grow the long lengths of cane that are required to cover the wirework (see next section).

During the March following planting, broadcast on the soil in a circle 18 in (45 cm) diameter round each plant:

1 oz (30 g) nitro-chalk
⅓ oz (10 g) sulphate of potash

or

2 oz (60 g) Phostrogen or Growmore

Similarly, at the end of May and June, broadcast:

⅔ oz (20 g) Nitro-chalk

In the succeeding years during March, broadcast on each sq yd (m^2) to a distance of 3 ft (90 cm), on both sides of the rows:

⅔ oz (24 g) nitro-chalk
⅓ oz (12 g) sulphate of potash

If the new canes do not grow to a satisfactory length, increase the amounts of nitrogenous fertiliser applied by 50 per cent. If the growth is too vigorous, reduce the amounts applied by 50 per cent.

Posts and Wires

The usual method of growing these fruits is on a framework of posts and wires. An alternative way is to grow them against a wall or fence, trained on wires threaded through vine eyes driven into the brick or woodwork. Bushes trained on walls with a northerly or easterly aspect will ripen their berries later than those with a southerly or westerly aspect.

The whole weight of the fruit and canes is carried by the wires, so the wires and the posts must be stouter than those required for training raspberries. The wire should be ten gauge (10 g) (3.15 mm). End posts should be 4 in × 4 in (10 cm × 10 cm) and 8½ ft (2.5 m) in length, driven 30 in (75 cm) into the ground.

Intermediate posts measuring 2 in × 2 in (5 × 5 cm) should be positioned 13 ft (4 m) apart in the row. Four wires should be loosely stapled to the intermediate posts at heights of 3 ft, 4 ft, 5 ft and 6 ft (90 cm, 120 cm, 150 cm and 180 cm). It is possible to reduce the number to three wires in very sheltered situations.

Methods for Training New Canes

Alternate Bay

The new canes, as they grow, should be trained and tied permanently to the wires in one direction and opposite to the direction in which the fruiting canes are tied (see Figure 2.2, bottom row). At any time after picking, it is easy to cut out at their bases the spent fruiting canes, leaving the wires free for training the new canes in the following year. The advantages of this method are that the new canes are separate from the old canes and less likely to be infected by diseases coming from the latter; the new canes are less likely to be damaged or split if they are tied to the wires whilst they are still supple; the training is easier to do and takes less time. As the fruiting canes are more crowded together compared with other methods, the yield of fruit may be slightly reduced.

New Cane on Top Wire

The new canes, as they grow in the spring, should be tied loosely in a bunch together and trained vertically until they reach the top wire (see Figure 2.2, top row). They should then be divided into two halves. One half should be trained to the right and the other half to the left along the top wire. During the following winter the canes should be taken off the top wire and can be trained on the lower three wires in their fruiting positions by a variety of systems. This method keeps the new canes separate from and above the fruiting canes, and they should therefore be less susceptible to disease infection. However, this involves more work than the previous method and the canes are more likely to be damaged whilst being moved from the top of the lower wires.

New Cane on the Ground

The new canes are divided into two halves and trained either way under the fruiting canes and on the ground (see Figure 2.2, centre row). The canes are tucked into both sides and kept in place with 2 ft (60 cm) long wires or wooden pegs. During the winter, after the spent fruiting canes have been cut out, the new cane is lifted off the ground preferably during March, and trained on the wires. The disadvantages of this system are that disease spores are washed down by rain from the fruiting cane onto the new canes, and some damage will be caused when the new canes are trained on the wires during the winter. The advantages are that during the summer a minimal amount of work is involved in training the new canes, and new canes lying on the ground during the winter are said to be less liable to be damaged by low winter temperatures.

Methods for Training Canes in their Fruiting Positions

Each of the following methods is equally adaptable to furnishing all the space on the wires, or half the wires in the alternate bay system.

Fan

The canes should be trained in the shape of the ribs of a fan and tied at every point at which they cross the wires (see Figure 2.2, left-hand column). This is the most time-consuming method but should produce the largest quantity of fruit.

Roping

When there are four wires, the canes should be divided into eight equal lots (see Figure 2.2, right-hand column). Each lot should be tied at intervals of 18 in (45 cm) in bunches, along one of the wires. Roping is the easiest training method but it produces the lowest yield of fruit.

Weaving

If this method is to be successful, it is essential that the wires remain taut. The canes should be divided into two halves. One half should be trained one way along the wires and the other half in the other direction (see Figure 2.2, centre column). Each cane should be trained first over the top wire and then under the bottom wire in succession. Each cane shold be trained 3-4 in (7-10 cm) along the wires and from the previous cane. A minimum

Figure 2.2: Blackberry and Hybrid Berry Training Systems

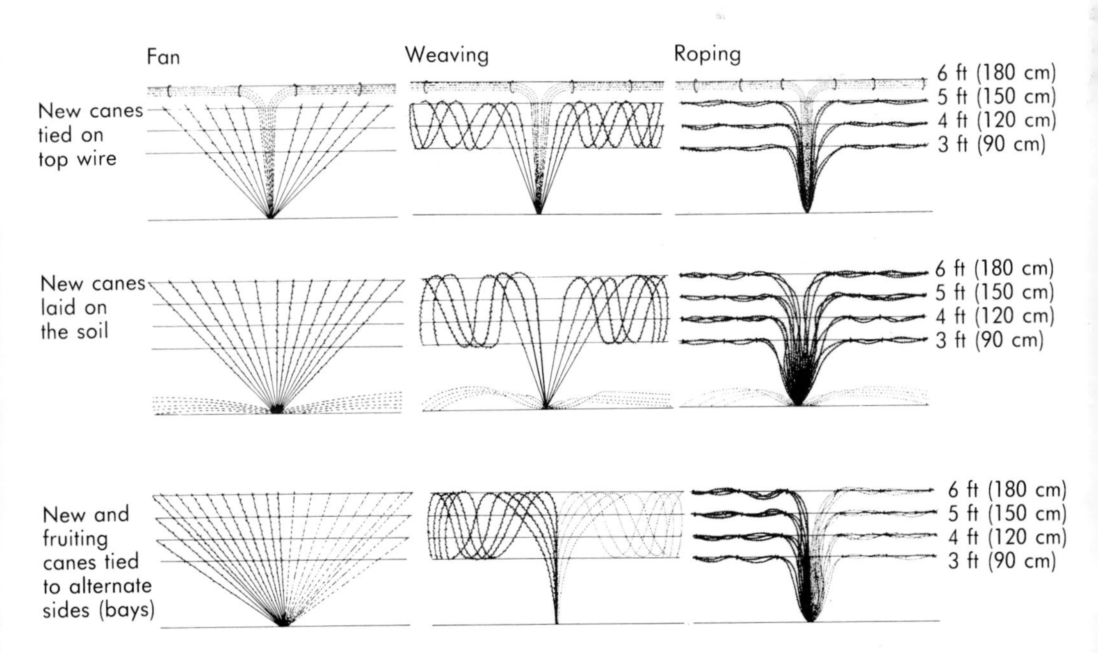

of two ties should be made to each cane, where each cane first passes over the top wire and where the cane passes last over or under a wire.

Random

The canes should be trained in a random pattern as long as they are fairly evenly distributed over the wires. When for any reason the canes are severely branched, this is the only practical way of training. Although the method looks untidy it is quite effective.

Table Top

The new canes, as soon as they are 3 ft 6 in long, should be trained in the opposite direction from the fruiting canes and led alternately up and under the cross-strings and at an equal distance from each other (see Figure 2.3). They should be left to bear fruit in this position in the following year.

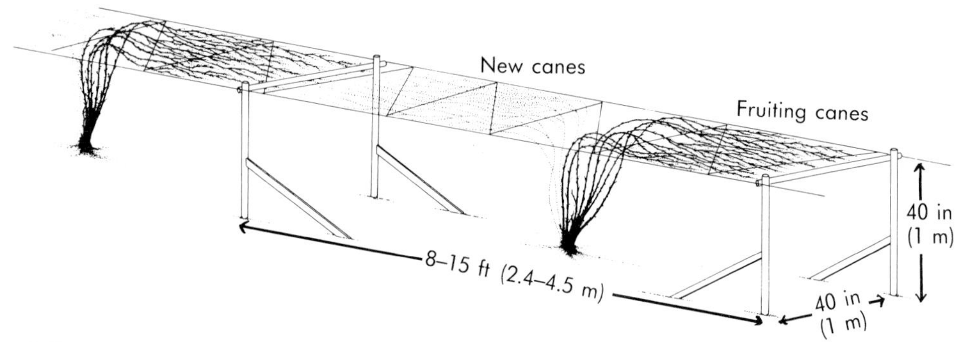

Figure 2.3: Blackberry and Hybrid Berry Table Top Training System with Alternate Sides. The uprights are 4 × 4 in (10 × 10 cm); cross and support struts 3 × 2 in (7.5 × 5 cm); the wire is 10 g (3.15 mm)

Weed Control

These berried fruits are remarkably resistant to applications of herbicide to control weeds. Gardeners should not hesitate to apply Casoron G in February or early March (UK) and Weedex in January or February (UK) to prevent the germination of annual weeds:

Light soils – one sachet of Weedex in 2 gallons (9 l) of water to 50 sq yds (42 m^2)
Heavy soils – one sachet of Weedex in 2 gallons (9 l) of water to 33 sq yds (28 m^2)

Casoron G should not be applied until the bushes are one year old, at the rate of ⅓ oz per sq yd (12 g/m^2), in February or early March (UK).

Pruning

Before planting, the canes should be cut to a length of 12 in (30 cm). In May or June (in the UK), when the new shoots are growing strongly, the old cane should be cut down to ground level to prevent flowering and fruit production, and to direct all the efforts of the plant into strong cane production. During the second and subsequent winters, all the spent fruiting canes should be pruned back to the stool (ground level) together with any new canes that are weak or broken. It is traditional to leave pruning until March as it is considered that the old canes give the new canes some protection from low winter temperatures. However, there is little evidence to support this theory.

Cane Numbers and Thinning

The weight of fruit that will be picked is closely related to the numbers of new canes that it is possible to retain and tie to the wires. The fan method of training can be expected to produce the heaviest weight of fruit because the canes are evenly spaced out to catch the greatest amount of sunlight. The canes should be tied 3–4 in (7.5-10 cm) apart to the top wire. A minimum of 24 canes per bush for all methods of training are required to fully furnish the wires. When the new canes are tied to the top wire or trained on the soil a small number of canes, in addition to these, should be retained to allow for breakages that occur whilst the canes are transferred to their fruiting positions. Canes in excess of these numbers should be cut out in May and June (UK) whilst they are still small. For the weaving, roping and tabletop methods of training, similar numbers of canes should be tied to the wires.

Watering

Blackberries and hybrid berries should be given up to 9 gallons per sq yd (50 l/m^2) in late June or early July (UK) to saturate the soil with water before the berries ripen.

Pests and Diseases

The raspberry beetle is the only pest, and botrytis grey mould the only disease against which it is necessary to apply an insecticide and fungicide each year. Blackberry purple blotch, when it occurs, should be controlled with routine applications of fungicide.

Propagation

Provided that the black or hybrid berry bushes have been cropping well and the leaves have a healthy appearance, they may be propagated either by layering or leaf and bud cuttings. Layering is the easiest method and is carried out in the open during July and August (in the UK). A small spade or trowel should be pushed into the soil to a depth of 6 in (15 cm) at an

Table 2.1: Blackberry and Hybrid Berry Protection Chart

Problem	Chemical	Key	Product	March	April	May	June	July	Aug.	Sept.
Aphids	Dimethoate	1	Boots Greenfly & Blackfly Killer							
		6	Murphy Systemic Insecticide							
	Fenitrothion	7	PBI Fenitrothion							
	Malathion	7	Malathion Greenfly Killer							
		6	Murphy Liquid Malathion							
	Permethrin	7	Bio Sprayday							
		2	Fisons Whitefly & Caterpillar Killer							
		3	Picket							
	Pirimiphos-methyl	3	Sybol 2							
Blackberry Purple Blotch	Copper	5	Comac Bordeaux Plus							
		6	Murphy Liquid Copper Fungicide							
	Benomyl	3	Benlate plus Activex							
	Carbendazim	1	Boots Systemic Fungicide							
	Thiophanate-methyl	4	Fungus Fighter							
		6	Murphy Systemic Fungicide							
Botrytis Grey Mould	Benomyl	3	Benlate plus Activex							
	Carbendazim	1	Boots Garden Fungicide							
	Thiophanate-methyl	4	Fungus Fighter							
		6	Murphy Systemic Fungicide							
Cane Spot	Benomyl	3	Benlate Plus Activex							
Cane Botrytis	Carbendazim	1	Boots Garden Fungicide							

Pest/Disease	Chemical		Product
Spur Blight	Thiophanate-methyl	4	Fungus Fighter
		6	Murphy Systemic Fungicide
Green Capsid Bug	Malathion	7	Malathion Greenfly Killer
		6	Murphy Liquid Malathion
	Fenitrothion	7	PBI Fenitrothion
	Permethrin	7	Bio Sprayday
		2	Fisons Whitefly & Caterpillar Killer
		3	Picket
	Pirimiphos-methyl	3	Sybol 2
Raspberry Beetle	Fenitrothion	7	PBI Fenitrothion
	Malathion	7	Malathion Greenfly Killer
		6	Murphy Liquid Malathion
	Rotenone	7	Liquid Derris
Red Spider Mite	Malathion	7	Malathion Greenfly Killer
		6	Murphy Liquid Malathion
	Pirimiphos-methyl	3	Sybol 2
	Dimethoate	1	Boots Greenfly & Blackfly Killer
		6	Murphy Systemic Insecticide
Rust	Propiconazole	6	Murphy Tumbleblite

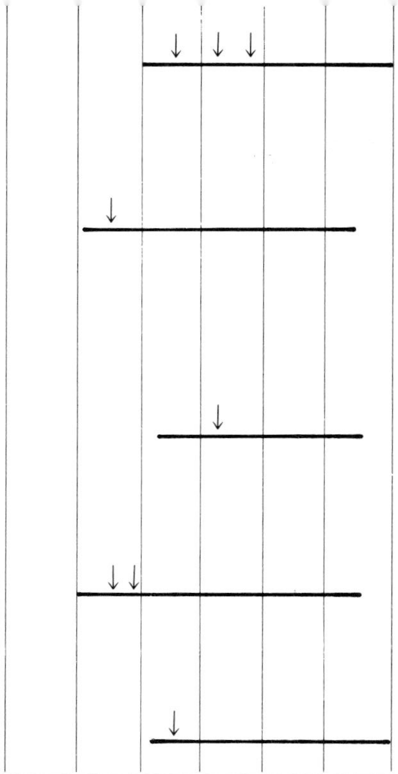

Key: ——— Infestation periods ↓ Timing of pesticide application

1 The Boots Company PLC, Nottingham.
2 Fisons PLC, Paper Mill Lane, Ipswich.
3 ICI Products, Woolmead House, Farnham, Surrey.
4 May & Baker, Regent House, Hubert Road, Brentwood.
5 McKecknie Chemicals Ltd, PO Box 4, Widnes, Cheshire.
6 Murphy Chemical Co., Latchmore Court, Brand Street, Hitchin, Herts.
7 Pan Britannica Industries, Waltham Cross, Herts.

angle of 45° away from the parent plant and where the tip can be easily positioned (see Figure 2.4(A)). The tool should be levered upwards so that the tip of the cane can be pushed into the slit that has been formed in the soil. The tool should be withdrawn; the soil then falls back on the tip and should be firmed with the sole of the boot. During the months following, one or more white shoots together with their roots will develop from the buried tip; these should not be lifted until the following March. They could then be planted out in their fruiting position or grown on 12 in (30 cm) apart in a short row for another year into strong one-year-old plants. Tips will root in soil that received an application of Weedex herbicide during the previous winter. Tips may also be rooted into 4½ in (11 cm) pots filled with compost and buried to their rims in the soil near to the plant that is to be used for propagation (see Figure 2.4(B)).

Figure 2.4: Blackberry and Hybrid Berry Propagation

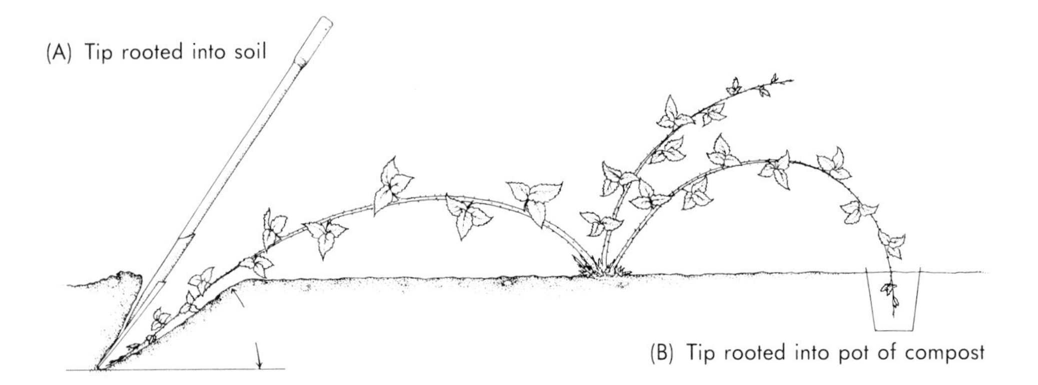

Leaf and bud cuttings may be taken and rooted at any time during the summer from June onwards (UK). A length of the new season's cane should be removed from the parent plant and cut into pieces 2 in (5 cm) long with one leaf and bud (see Figure 2.5). The cuts should be immediately above and 2 in (5 cm) below the buds. They may be rooted in a propagating case or under mist with a bottom temperature of 70°F (20°C) in a greenhouse or in a cold frame. The cuttings root in 6-8 weeks, after which they should be potted on, hardened off and either lined out in a short row or grown on in suitable sized pots.

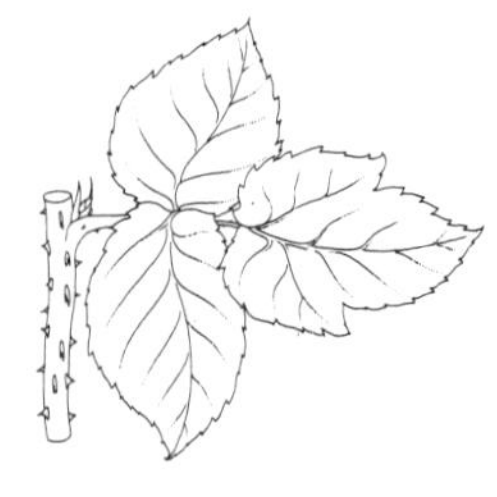

Figure 2.5: Blackberry and Hybrid Berry Leaf and Bud Cutting

Chapter 3 Blackcurrants

Introduction

The traditional blackcurrant varieties such as 'Baldwin' and 'Wellington XXX' are no longer worth growing because they have been made obsolescent by the introduction of heavier cropping varieties. Many of these are resistant to mildew and to damage by spring frosts and cold winds. Bushes of the heavier cropping 'Ben Lomond' are now readily available from nurserymen. 'Ben Sarek', an even heavier cropping dwarf variety, has just been introduced and one-year-old bushes should be available for purchase in the autumn of 1985. Bushes grow in zones of hardiness 5-8.

Varieties

'Ben Lomond'. (SCRI)

This is a mid-season variety that ripens its fruit during the last few days of July (in the UK). The berries are very large, ripen evenly, have tough skins and a high vitamin C content. Jam, juice and other products made with the fruit are of very good colour but have a slightly weaker flavour than traditional varieties. The bush is upright, moderately vigorous, compact and requires a minimum of pruning. 'Ben Lomond' is late-flowering and has resistance to cold. However, it is susceptible to mildew. The flavour is good.

'Ben More' (SCRI)

'Ben More' ripens its fruit seven to ten days after 'Ben Lomond' and was introduced because it is late-flowering and resistant to frost, but it has not fulfilled its earlier promise as far as yields are concerned. It has a vigorous upright habit of growth and is resistant to mildew. The berries are large with a low vitamin C content. The juice is of good colour and acidic flavour.

'Ben Nevis' (SCRI)

'Ben Nevis' is very similar to 'Ben Lomond' but is more vigorous with more spreading branches. The berries ripen three to seven days earlier and are not quite so large as those of 'Ben Lomond'. Two reasons for growing 'Ben Nevis' rather than 'Ben Lomond' would be to pick fruit earlier or because the soil is poor. The flavour is good.

'Ben Sarek' (SCRI)

'Ben Sarek' has been introduced specifically for growing in private gardens because it forms a small compact bush of medium vigour which rarely grows more than 40 in (100 cm) in height (see Plate 7). The berries are very large and borne on short strigs that are easy to pick. In trials, 'Ben Sarek' has given yields that exceed those of all other blackcurrant varieties by 30 to 50 per cent, so much so that the weight of the crop frequently bears the branches down to the ground. It is tolerant to frost and cold injury and resistant to mildew. On soils of average fertility it could be planted as close as 4 ft × 4 ft (120 cm × 120 cm). The flavour is fairly good.

'Jet' (EMRS)

'Jet' is only a moderate yielding variety but its good attributes are that it extends the blackcurrant season by two to three weeks and is late-flowering, usually escaping spring frost damage. The berries are small but are easy to pick because as many as 20 are borne on each strig and when ripe will hang on the bushes for up to 14 days without deteriorating. The bushes are large, spreading and not very susceptible to mildew. The flavour is described as bland.

'Tsema' (The Netherlands)

'Tsema' is an early-ripening variety and, pruned in the usual way, the weight of fruit will be lower than that of the varieties already described because of its very spreading habit. The yield can be improved if the lower branches are not cut off but supported above the ground by strings tied to a stake in the middle of each bush. It is susceptible to mildew and has little frost resistance. The berries are small and borne on long strigs. The flavour is fairly good.

Types of Bush

Only bushes that have been certified by the Ministry of Agriculture or Department of Agriculture for Scotland should be bought. One-, two- and three-year-old bushes should be available for purchase.

Certified one-year-old bushes were available for the first time in 1983. Bushes should have a minimum of two branches and should be over 18 in (45 cm) in length, together with a good fibrous root system. They are very good value for money but they will take longer than two-year-old bushes to come into full bearing.

Two-year-old bushes should have between four and six branches each over 24 in (60 cm) and a large fibrous root system. Though more expensive than one-year-old bushes they are good value for money.

Three-year-old bushes are not good value for money because they are rarely supplied with a large enough root system to correspond in size with the six or more 3 ft (90 cm) branches. They suffer too large a check on transplanting compared with younger bushes and it is unlikely that the amount of fruit picked will be any larger than that borne by the younger bushes. They are not a good buy.

Planting Distances

Bushes should not be planted in rows less than 5 ft (1.5 m) wide. At this distance there is scarcely sufficient room to carry out spraying, hoeing and picking. Six feet (1.8 m) would be a better distance for most varieties, though it appears that row distance for 'Ben Sarek' should be 5 ft (1.5 m).

The distance between the bushes in the row is less important and should be 4-5 ft (1.2-1.5 m), but 'Ben Sarek' need not be more than 4 ft (1.2 m). If it should be necessary to grow the maximum quantity of fruit as quickly as possible, the distance that bushes should be planted in the rows is 2-2½ft (60-75 cm). This will double the yield of fruit for the first four years but after this it will be similar to that from bushes planted at the wider distances.

Planting

In the United Kingdom, it is important that bushes should be planted before Christmas because blackcurrants come into growth early in the New Year. If planting is delayed, many small white roots are destroyed and growth will not be as good as it otherwise would have been. A hole should be dug wide enough for the roots to be spread out and deep enough for the fork, where the branches divide, to be just covered after the soil has been replaced and firmed over the roots. On shallow soils it can sometimes be difficult to do this without the lower roots being planted in the sub-soil. In this case the whole bush should be planted at an angle of 45° in a shallower hole.

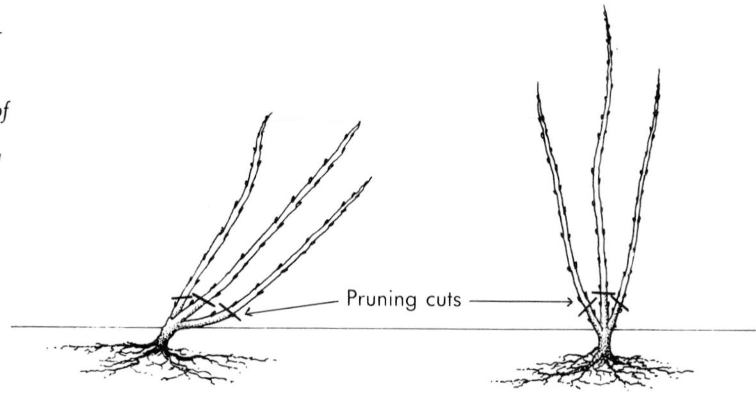

Figure 3.1: Blackcurrant Bushes – Normal Planting and at an Angle of 45°. The bushes should be pruned immediately, as shown

Pruning

Immediately after planting, each branch should be pruned hard back to two buds showing above soil level. *None of the branches should be left in place to bear fruit for they will compete with the new shoots for nutrients and water.* Such competition will delay the development of fully grown bushes.

At the end of the first year, if one- or two-year-old bushes were planted, for every branch that was pruned back, two or more branches each at least 30 in (75 cm) should have grown in its place, with possibly, in addition, one or two weaker shoots. Pruning should consist of cutting out these weaker shoots and cutting back one strong shoot to the base of the bush. The purpose behind this pruning is to encourage the production of strong basal shoots in a bush that, for its size, could be bearing too heavy a crop of fruit.

In subsequent years, all branches that are growing out from and at an angle of less than 45°, should be cut off. If they are left in place, the weight of fruit they bear will bring them down to the ground where the fruit will get dirty, become liable to infection by botrytis or be eaten by slugs.

Blackcurrants bear more and better quality fruit on the previous seasons growths. When the bushes are fully grown, they should be pruned every year to prevent the accumulation of too much old wood and encourage the production of strong new growths. With this objective, prune back three or four of the oldest branches to strong new growths or, in their absence, to the base of the bush. (See Figure 3.2.)

Fertiliser Application

At one time it was the practice to recommend that quite large quantities of nitrogenous fertiliser be applied to blackcurrant bushes each year. However, the application of herbicides has eliminated the growth of weeds; they no longer compete with

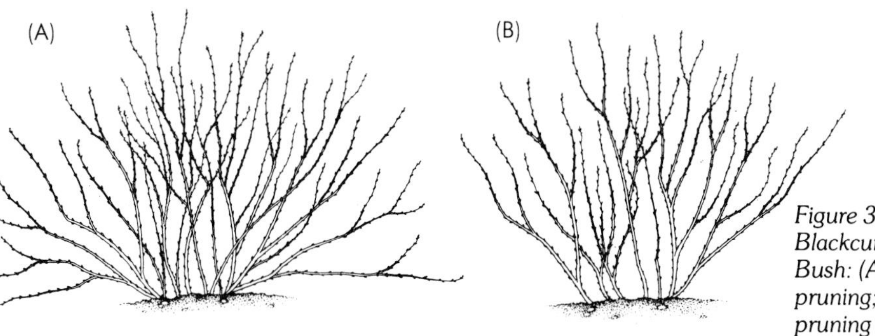

Figure 3.2: Mature Blackcurrant Bush: (A) before pruning; (B) after pruning

the bushes for nutrients and water and, as cultivation is no longer necessary, the undisturbed roots have more soil to exploit. With this method of management less nitrogenous fertiliser should be applied.

In the United Kingdom, immediately after planting or at the end of March, whichever is the later, broadcast in a circle of 18 in (45 cm) round each bush.

1 oz (30 g) nitro-chalk
⅓ oz (10 g) sulphate of potash

or

2 oz (60 g) Growmore

or

2 oz (60 g) Phostrogen

In the years following, until the branches of the bushes meet in the row, during March (UK), broadcast the above amounts of the fertiliser selected over each sq yd (m^2), 3 ft (1 m) either side of the row.

When the bushes meet in the row, the amounts of fertiliser broadcast should be reduced to:

⅓ oz per sq yd (12 g/m^2) nitro-chalk
⅓ oz per sq yd (12 g/m^2) sulphate of potash

In every fifth year, in addition to above, broadcast:

2 oz per sq yd (70 g/m^2) superphosphate of lime

or

1 oz per sq yd (36 g/m^2) Growmore

or

1 oz per sq yd (36 g/m^2) Phostrogen

The amount of nitrogen applied should be adjusted up or down if the lengths of the new growths are significantly shorter or longer respectively than 2½ ft (75 cm).

Weed Control

Weedex should be applied to the soil during February or after planting, whichever is the later (in the UK). In the years following, the Weedex should be applied each February, at a rate of application of:

Light soils – one sachet per gallon (4.5 l) applied to 50 sq yds (42 m^2)
Heavy soils – one sachet per gallon (4.5 l) applied to 33 sq yds (28 m^2)

If there are any established weeds or seedlings present, they should be killed by an application of one sachet of Weedol in 1 gallon (4.5 l) of water applied to 20 sq yds (17 m^2). Where there are no weeds present, there is no point in applying Weedol to clean ground. This herbicide should be applied with either a dribble bar or a coarse spray so that none falls on the blackcurrant buds.

Watering

Blackcurrants should be kept well supplied with water and not allowed to suffer from drought at any time during the summer period May to August. Watering can increase berry size, double the weight of fruit picked and ensure that the new shoots grow satisfactorily.

Picking

Provided blackcurrants are protected from birds, the berries should be left to hang on the bushes until 99 per cent of the crop is ripe. All the varieties listed, with the exception of 'Jet', should be picked at this stage, otherwise the berries will drop to the ground. 'Jet' will remain hanging on the bushes for about two weeks whilst fully ripe.

If the bushes are not netted against birds, the crop should be gathered by picking the berries individually as they ripen before the birds get them – a somewhat tedious task!

The harvesting of the fully ripe crop can be made easier and 'mechanised'. All that is required is a sheet of polythene that covers the soil under half of the bush, and a beating stick – a piece of hardwood 18 in (45 cm) long and 1½ in (4 cm) in diameter. Two or three branches should be taken in the left hand and pulled over the sheet and beaten between the hand and bush centre. The vibrations transmitted to the fruit trusses make the berries drop onto the sheet. The fruit should be easily removed from below the bush on the sheet and emptied into a bucket. The other half of the bush should then be harvested.

There will be a fair amount of extraneous debris mixed with the fruit such as leaves, fruit stalks, snails and earwigs. These should be removed by passing the berries over a sieve, blowing air over the fruit with a hair drier and/or flotation in water. It is inadvisable to use water if the fruit is to be frozen.

Another, easier, method of picking sometimes used is that of cutting out the branches that bear the fruit and taking them indoors where the berries can be removed sitting down. A small amount of fruit will remain on the bush and this should be picked *in situ*. If this method is used care should be taken to leave sufficient new shoots on the bushes to give a satisfactory crop in the following year.

Pests and Diseases

The bushes should be kept under careful observation for the presence of big bud mite and reversion virus. Reverted bushes should be dug up and burned; swollen buds should be picked off and burned. Routine sprays should be applied to control aphids, big bud mite, leaf spot and mildew.

Propagation

A warning has already been given about the dangers of propagating unhealthy bushes. This work should only be undertaken if the bushes from which the cuttings are to be taken come from bushes *that are not more than four years old*, were certified at the time of planting, and are not affected by or growing near to bushes that are infected by big bud mite or reversion virus.

In October or November (in the UK), immediately after leaf fall, any one-year-old wood that would be taken out of the bushes in the normal way should be pruned and may be used to make hardwood cuttings. Traditionally, the one-year-old wood is cut into 8 in (20 cm) lengths but the thin tip-most portions are discarded. Cuttings that are as short as 3 in (7.5 cm) long are equally satisfactory and are easier to lift twelve months later.

The cuttings should be planted immediately 6 in (15 cm) apart in one row or in rows 24 in (60 cm) apart. The soil should have been prepared by digging, raking and rolling or tramping until it is as firm as a rough seed-bed for peas or beans. The cuttings should be pushed vertically in the soil until the top-most bud is just showing. If there is no ground immediately available, the cuttings (in bundles) can be stored buried in sand in the open

Table 3.1: Blackcurrant Protection Chart

Problem	Chemical	Key	Product	April	May	June	July	Aug.	Sept.	Oct.
American Gooseberry Mildew	Bupirimate and Triforine	3	Nimrod T							
	Sulphur	5	Comac Cutonic Sulphur Flowable							
Aphids	Dimethoate	1	Boots Greenfly & Blackfly Killer							
		6	Murphy Systemic Insecticide							
	Fenitrothion	7	PBI Fenitrothion							
	Malathion	7	Malathion Greenfly Killer							
		6	Murphy Liquid Malathion							
	Permethrin	7	Bio Sprayday							
		2	Fisons Whitefly & Caterpillar Killer							
		3	Picket							
	Pirimiphos-methyl	3	Sybol 2							
Blackcurrant Gall Mite	Benomyl	3	Benlate plus Activex							
	Carbendazim	1	Boots Garden Fungicide							
	Thiophanate-methyl	4	Fungus Fighter							
		6	Murphy Systemic Fungicide							
	Sulphur	5	Comac Cutonic Sulphur Flowable							
Blackcurrant Leaf Curling Midge	Dimethoate	1	Boots Greenfly & Blackfly Killer							
		6	Murphy's Systemic Insecticide							

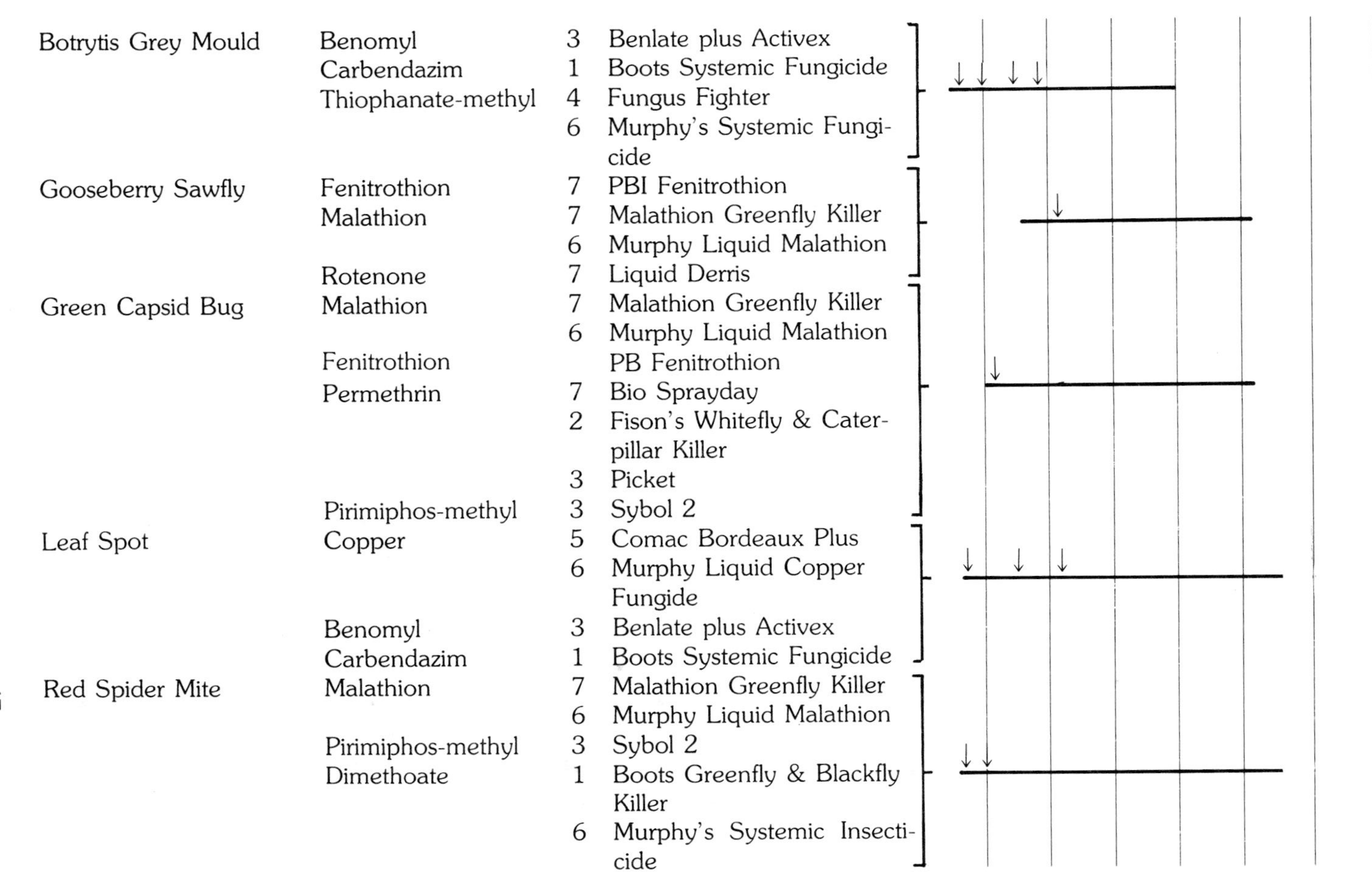

Pest/Disease	Chemical		Product
Botrytis Grey Mould	Benomyl	3	Benlate plus Activex
	Carbendazim	1	Boots Systemic Fungicide
	Thiophanate-methyl	4	Fungus Fighter
		6	Murphy's Systemic Fungicide
Gooseberry Sawfly	Fenitrothion	7	PBI Fenitrothion
	Malathion	7	Malathion Greenfly Killer
		6	Murphy Liquid Malathion
	Rotenone	7	Liquid Derris
Green Capsid Bug	Malathion	7	Malathion Greenfly Killer
		6	Murphy Liquid Malathion
	Fenitrothion		PB Fenitrothion
	Permethrin	7	Bio Sprayday
		2	Fison's Whitefly & Caterpillar Killer
		3	Picket
	Pirimiphos-methyl	3	Sybol 2
Leaf Spot	Copper	5	Comac Bordeaux Plus
		6	Murphy Liquid Copper Fungide
	Benomyl	3	Benlate plus Activex
	Carbendazim	1	Boots Systemic Fungicide
Red Spider Mite	Malathion	7	Malathion Greenfly Killer
		6	Murphy Liquid Malathion
	Pirimiphos-methyl	3	Sybol 2
	Dimethoate	1	Boots Greenfly & Blackfly Killer
		6	Murphy's Systemic Insecticide

Problem	Chemical	Key	Product	April	May	June	July	Aug.	Sept.	Oct.
Rust	Copper	5	Comac Bordeaux Plus							
		6	Murphy Liquid Copper Fungicide							
	Propiconazole	6	Murphy Tumbleblite							
Slugs and Snails	Methiocarb	7	Slug Gard							
Tortrix Moth	Fenitrothion	7	PBI Fenitrothion							
	Malathion	7	Malathion Greenfly Killer							
		6	Murphy Liquid Malathion							
	Permethrin	3	Picket							
Winter Moth	Fenitrothion	7	PBI Fenitrothion							
	Permethrin	7	Bio Spray Day							
		2	Fisons Whitefly & Caterpillar Killer							
		3	Picket							
	Pirimiphos-methyl	3	Sybol 2							

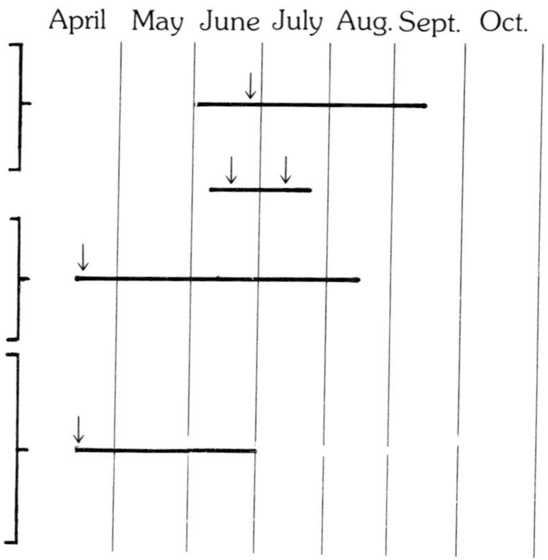

Key ——— = Infestation periods ↓ = Timing of pesticide application

1 The Boots Company PLC, Nottingham.
2 Fisons PLC, Paper Mill Lane, Ipswich.
3 ICI Products, Woolmead House, Farnham, Surrey.
4 May & Baker, Regent House, Hubert Road, Brentwood, Essex.
5 McKecknie Chemicals Ltd, PO Box 4, Widnes, Cheshire.
6 Murphy Chemical Co., Latchmore Court, Brand Street, Hitchin, Herts.
7 Pan Britannica Industries, Waltham Cross, Herts.

until they are required; if the cuttings are stored much later than December (UK), they develop roots and these break off whilst planting.

The soil should receive an application of Weedex in exactly the same way as fruiting bushes to keep the soil weed free (see p. 44). The top-most buds should grow out towards the end of April. At this time sprinkle along the row(s) 1 oz per sq yd (36 g/m^2) of Growmore fertiliser and at the end of June similarly apply ½ oz per sq yd (18 g/m^2) of nitro-chalk. The new shoots of the cuttings are likely to be attacked by the same pests and diseases as the fruiting bushes and should receive the same spray programme.

Chapter 4
Blueberries

Introduction

Blueberry fruit sold in the United Kingdom is almost certain to have come from the USA by air freight. In that country, the native species *Vaccinium corymbosum* and *V. australe* grow and from them the cultivated varieties have been derived by breeding. In the UK a handful of gardeners grow blueberries for their fruit and there is one successful commercial plantation in Dorset. In the autumn, the bushes exhibit attractive highly coloured red and yellow tints.

A considerable area of blueberry plantations are cultivated for fruit production on Luneburg Heath in northern Germany at latitude 53°N which also passes through the centre of England between Manchester and Birmingham. This would indicate that there should be no climatic reasons why blueberries could not grow successfully in the United Kingdom (i.e. zones of hardiness 5-8). Since 1967, an investigation into blueberry production at the Scottish Crop Research Institute has demonstrated that the crop can be grown successfully; discovered the various problems that are likely to be encountered; and recommended the most suitable varieties for planting. It has been demonstrated that blueberries are most demanding in the soil conditions they require, the bushes are very slow growing and their cultivation requires patience and attention to detail. The bushes should not be allowed to bear fruit for the two years after planting and a further seven years pass before they come into full production when yields of 6-8 lb (2.5-3.5 kg) of berries should be expected; they are extremely long lived and should go on cropping for 20 years or more.

Varieties

The season of ripening for blueberries is normally, in the south of England, from the last week of July until the end of September and in Scotland, three weeks later. The season can vary from year to year by as much as three weeks in England and two weeks in Scotland. The ripening period for individual varieties is approximately three weeks. A small number of variety trials have been carried out at the Scottish Crop Research Institute and by the Agricultural Development Advisory Service in England. These would suggest that the varieties that are worth growing are 'Earliblue', primarily because it is early-ripening, the mid-season varieties 'Bluecrop' and 'Berkeley', or 'Ivanhoe' which is resistant to the fungal disease *Godronia cassandrae* (canker) and one of the late varieties, 'Coville' or 'Heerma 1', the latter being resistant to *Godronia cassandrae*. Blueberries are self-fertile, but at least two varieties should be planted as cross-pollination gives an improved set of seeds and larger berries.

'Berkeley'

Bush vigorous and spreading, high yielding but more variable in cropping than 'Bluecrop'. The berries are very large, light blue, firm and well flavoured. Ripening mid-season.

'Bluecrop'

This is probably the one variety that should always be selected for planting. The bushes grow vigorously with an upright habit. They bear heavy crops of large light blue berries of good flavour. In America it is said to be more resistant to drought than other varieties. Ripening mid-season. (See Plate 8.)

'Bluehaven'

The bush is moderately vigorous, upright with a compact habit. The berries are large, firm, light blue in colour and of very good flavour. Ripening mid-season.

'Collins'

One fault of this variety is its susceptibility to *Godronia cassandrae* but it has the advantage of ripening one week earlier than the other mid-season varieties and bearing very highly coloured leaves in the autumn. It gives a moderate yield of large, light blue excellently-flavoured berries.

'Coville'

The bush is vigorous, spreading and high yielding. The berries are very large, light blue in colour, and of good flavour. It is late-ripening.

'Earliblue'

This variety is a vigorous, upright grower but the yield is only moderate. The berries are large, light blue and of good flavour. The main reason for growing this variety would be to extend the season as it ripens two weeks before other varieties.

'Heerma 1'

This is a variety that was bred in Germany and is extremely resistant to *Godronia cassandrae*. The bush is vigorous and high yielding but the berries are only of moderate size and flavour. It is likely to grow well in most situations and is late-ripening.

'Herbert'

This variety is considered to be better flavoured than all the other varieties. The bush has a vigorous, upright habit. The berries are very large with a medium blue colour. It is mid-season ripening.

'Ivanhoe'

This is the most resistant of the American varieties to *Godronia cassandrae*. It bears good crops of very small darker blue-coloured berries. It ripens in mid-season.

Growing Conditions

The most important condition that blueberries require in order to grow and crop successfully is acid soil with a pH of 4.0 to 5.0. Very few garden soils are likely to be as acid as this. Even in areas where naturally acid soils occur, they have usually been intensively farmed and limed before being sold for house building. The majority of established garden soils are very alkaline because most garden books recommend an application of lime for almost every crop apart from those that are known to require acid soil conditions.

Blueberries require a moisture-retaining but free-draining soil that is at least 12 in (30 cm) deep. Such soils, when rubbed between the fingers, should have a coarse, gritty feel due to the presence of sand or small stones. Heavy soils that have a smooth feel, and when turned over with a spade leave a polished surface, should be avoided unless they have a very high organic matter content – 10-12 per cent – and coarse grit added to them. Even the lighter sandy and gravelly soils should have a high organic matter content of 5-6 per cent. It is essential if bushes are to grow and crop satisfactorily to apply lime-free water to the bushes in times of drought.

Climatic conditions in those parts of the United Kingdom where other soft fruits are grown should be suitable for blueberries. They require shelter from the prevailing winds. When sited on low-lying land or in frost pockets, blossoms can be killed by spring frosts in April and May.

The soils that are situated above the chalk of south-east England or the limestone of the Pennines need not necessarily be alkaline. However, if particles of the underlying chalk or limestone have become incorporated in the topsoil, it would be uneconomic to acidify such a soil by applying sulphur or peat. In these areas, it is likely that the mains water would be 'hard', contain lime and be unsuitable for applying to blueberries.

Soil Acidity

Unless the soil is acidified and reduced to a pH of between 4.0 and 5.0, the blueberry bushes will die. The soil could be sent to a testing laboratory with a request that the sulphur requirement be determined to bring the soil down to pH 4.5. Alternatively, the pH can be measured with a colour soil-testing kit or electronic soil meter. Flowers of sulphur, peat or sawdust should be added and thoroughly mixed by forking or rotavating into the top 8 in (20 cm) of soil in accordance with Table 4.1. The sulphur should be applied twelve months before planting and if this is not possible, mix half a gallon (2 l) of peat with the soil in the planting hole. Sawdust is not as effective as peat in reducing the pH; therefore it would be advisable, particularly where heavy applications are required, to use equal parts of each. This would reduce the cost as would combined applications of sulphur and peat or sawdust. It is also possible that applications of more than 12 in (30 cm) of sawdust could lead to adverse soil conditions. Soils that are low in organic matter should receive as much peat or sawdust as circumstances allow to increase the moisture-holding capacity of the soil.

Table 4.1: Recommended Applications of Flowers of Sulphur, Peat and Sawdust to Increase Soil Acidity

Soil	Flower of Sulphur				Peat				Sawdust			
	Light Soil		Medium Soil		Light Soil		Medium Soil		Light Soil		Medium Soil	
Ph	sq yd	M^2	sq yd	M^2	sq yd	M^2	sq yd	M^2	sq yd	M^2	sq yd	M^2
5.5	1 oz	36 g	2 oz	72 g	4 in	10 cm	8 in	20 cm	4 in	10 cm	8 in	10 cm
6.5	2 oz	72 g	4 oz	144 g	8 in	20 cm	16 in	40 cm	8 in	20 cm	16 in	20 cm
7.5	3 oz	108 g	6 oz	216 g	12 in	30 cm	24 in	60 cm	12 in	30 cm	24 in	30 cm

The soil should be re-tested two years afterwards and, if the acidity has been increased materially beyond pH 4, the growth of the bushes could be adversely affected, and hydrated lime should then be given as a top dressing. To raise the pH by one unit, 5 oz per sq yd (170 g/m^2) or 8 oz per sq yd (270 g/m^2) of hydrated lime should be added to light and medium soils respectively. In later years, if the leaves lose their fresh green colour before the time when autumn colours develop, the soil should first be re-tested to make certain that a rise or lowering of the pH is not the cause of the discoloration.

Planting Stock

Blueberry bushes are not eligible for certification, therefore to obtain good bushes, buy from a nurseryman who has a reputation for supplying good quality plants of all kinds. Bushes are certified in the USA and, occasionally, these are imported and offered for sale in the UK. Blueberries are difficult to propagate and this makes them expensive. They are usually offered for sale in pots or polythene containers. Buying two-year-old bushes is a reasonable compromise between price and size of bush.

Planting Distances

4-5 ft (1.2-1.5 m) between the bushes
5-6 ft (1.5-1.8m) between the rows

Soil Preparation

Blueberries develop a mat of very fine roots in the surface soil and it is important that these should be disturbed as little as possible. To ensure that the soil is completely free from perennial weeds before planting, carry out the methods recommended on page 5 for the elimination of these weeds. If this has not been done successfully, delay planting for a year whilst further weed eradication procedures are followed.

Planting

If the roots or root ball are dry they should be soaked in water before planting. Planting holes should be dug so that the bushes can be planted at the same depth as they were growing in the nursery or so that the root ball is level with the surface of the ground. The soil should be well firmed on the roots or round the root ball.

Pruning after Planting

Blueberry bushes should not be allowed to bear any fruit for two years after planting so that all their energies are concentrated on increasing their size. No pruning should be done other than that of cutting out damaged and diseased wood. Cropping should be prevented by rubbing out all the fat fruiting buds, leaving only vegetative buds to grow.

Manuring

Blueberry bushes require very small amounts of fertiliser; if large amounts are applied the roots are likely to be damaged, growth will be poor and the crop reduced. Fertilisers that contain lime or calcium should not be used. It can be assumed with safety that garden soils which have been growing vegetables and fruit for a number of years contain excessive amounts of phosphorus and potassium and do not require additional fertiliser. New gardens or soils that have not been cropped previously with fruit or vegetables should have worked into them before planting:

⅔ oz per sq yd (24 g/m^2) sulphate of potash
1 oz per sq yd (36 g/m^2) superphosphate

Nitrogen could be applied to blueberries in the form of hoof and horn, ammonium nitrate, sulphate of ammonia or urea. Sulphate of ammonia is the cheapest and easier form to buy. The recommended rate of application to be made each March (in the UK) is:

½ oz per sq yd (18 g/m^2) sulphate of ammonia

Broadcast at this rate over the soil, 10 in (25 cm) beyond the spread of the branches. When the bushes are fully grown, broadcast the fertiliser over all the soil. This rate of application should be increased by a half or more if the bushes do not make sufficient new growth, provided the leaves are a healthy bright green colour.

At the same time, each year, similarly broadcast:

⅓ oz per sq yd (12 g/m^2) sulphate of potash

Weed Control

As the blueberry has a fine root system near the soil surface that should be disturbed as little as possible, annual weeds should be controlled by the application of Weedex, even though this is less effective on soils that have a high organic matter content or have had large quantities of peat or sawdust mixed with them. The Weedex should be applied as outlined on page 19. The rates of application are:

Light soils – one sachet of Weedex in 2 gallons (9 l) of water to 50 square yds (42 m^2)

Medium soils – one sachet to Weedex in 2 gallons (9 l) of water to 33 sq yds (28 m^2)

Casoron G is also safe to use amongst blueberries. It should be applied primarily for the spot treatment of certain perennial weeds. Although very expensive compared with Weedex, Casoron G may be used to control annual weeds as it is effective on high organic soils. Casoron G should be applied in February (in the UK). The rate of application for annual weeds is:

⅙ oz per sq yd (6 g/m^2)

Mulching

Weeds may also be controlled by covering the soil with a 2-3 in (5-7.5 cm) thickness of sawdust. If possible, wet sawdust, or sawdust that has begun to decompose, should be applied as new dry sawdust can easily be blown away by the wind. The sawdust gradually decomposes, and some of it becomes incorporated in the soil making, from time to time, small additional applications necessary. A mulch of tree bark could be applied instead of sawdust, though it would be more expensive.

A mulch of black polythene is as satisfactory as sawdust and sometimes leads to better growth of the blueberry bushes. However, polythene is expensive and can be unsightly, particularly when it is torn and flaps about in the wind. As the polythene has to remain in place for a long time the thickness should be at least 500 g and 1,000 g would be even better. An additional advantage of polythene and sawdust mulches is that they conserve moisture in the soil to the equivalent of 2 in (5 cm) of rain.

Intercropping

As blueberries are slow growing and take many years to occupy all the space allocated to them, there are no reasons why the soil between the rows of bushes should not be intercropped for a number of years. Suitable crops would be strawberries, lettuce, onions and radish. To do this, the centre strip of soil 3 ft (1 m) wide should not be acidified or covered with a mulch until after the growing of these crops has ceased.

Annual Pruning

The regular pruning of blueberries should start in the third winter after planting. The objective should be to maintain a balance between new shoots and older branches, retain a well shaped bush and remove all diseased and damaged growths. If bushes are pruned too severely the yield of fruit will be significantly reduced, though the individual berries will be larger and ripen earlier. Therefore a moderate system of pruning should be adopted. This consists of cutting out damaged, broken or diseased shoots or branches, back into sound wood. The low-lying branches of spreading varieties that are likely to be borne down onto the soil should be cut right out. It may be necessary to cut out some of the branches of erect growing varieties to prevent overcrowding, or to allow air and light into the bushes and make picking easier. The clusters of thin bushy wood that accumulate on the older branches should be removed by cutting these branches back to ground level or to a strong side branch. In mature bushes, this method removes about a fifth to a sixth of the old branches each year. If the bushes are not growing in a fruit cage, pruning should be left until the end of March (UK).

The vigour of bushes that are more than twelve years old may decline, making it difficult to follow this method of pruning and maintain sufficient young wood in the bushes. The bushes should be invigorated by pruning all the branches down to ground level; no fruit will be produced in the following summer but there should be many strong growths to bear a heavy crop in the year following. A small number of the weaker shoots should be cut out to prevent overcropping in this year and ensure that further new growths will appear.

Watering

Blueberries should never be allowed to suffer from lack of water and should be watered whenever necessary from May to August. Fully grown bushes require 4½ gallons per sq yd (25 l/m^2) each week either as rain or applied soft water, if they are to grow satisfactorily and bear large juicy berries.

Picking

The bushes should be picked over every four to five days. When fully ripe, the berries detach easily from their stalks. The method of picking should be to place each hand under a cluster of berries which should be lightly rolled between the thumb and fingers so that the ripe berries fall into the palms and are transferred to a basket. The fruit has a good shelf life and will keep at atmospheric temperature for a week or more; in a refrigerator the berries will keep for up to three weeks. They are also an excellent fruit to store by deep freezing.

Pests and Diseases

At present, blueberries have remained remarkably free from pests and diseases. In occasional years winter moth caterpillars feed on the leaves and tortrix moth caterpillars feed in the growing points of the shoots. The growing points are also sometimes infested with aphids.

The worst trouble that has affected the bushes is canker (*Godronia cassandrae*). Symptoms of this disease are brown areas surrounded by yellow to bright-red margins on the wood of the shoots and branches (see Plate 16). When girdled, the leaves wilt and the shoots or branches are killed. The flowers and swelling fruits can be infected by botrytis grey mould, so routine applications of fungicide should be made, particularly after spring frosts have occurred.

Propagation

Blueberries are propagated by soft and hard wood cuttings. They do not root easily, which is probably the reason why so few nurserymen have plants for sale and why they are so expensive. Propagation of blueberries could be very difficult for a gardener who is away from home during the day and unable to tend the cuttings.

Hardwood cuttings should be rooted in an outdoor cold frame of any size that is 12-16 in (30-45 cm) deep. The bottom should be covered with 1-2 in (2.5-5 cm) of coarse grit covered by a 6 in (15 cm) depth of a mixture of three parts of peat and one part of coarse sand. In March, the cuttings should be made from well ripened shoots that grew during the previous summer. Parts of the shoot that are unripened or bear fruit buds should not be used to make part of the cuttings. They should be made 4 in (10 cm) long with a sharp cut above and below the topmost and lowest buds respectively. Insert the cuttings two-thirds of their length into the compost in rows 4 in (10 cm) apart and 2 in (5 cm) apart in the rows. Lightly water afterwards and sufficient water should be applied at intervals to prevent the peat from drying out. Although the buds begin to grow out and the cuttings form calluses as soon as temperatures begin to rise, roots do not appear until mid-June, when the amount of water applied should be increased. Before this period, ventilation and shading should be carefully applied to prevent temperatures rising too high in the frame. After rooting has occurred, the cuttings should be hardened off and the frame cover removed. The rooted cuttings should be removed from the frame in the following March and, as they are rather too small to be planted in their permanent fruiting positions, either they should be lined out in nursery rows 12 in × 18 in (30 × 45 cm) or grown on in pots for one or two years.

Blueberries may also be propagated by soft wood cuttings taken from bushes in June when new lateral shoots are more than 6 in (15 cm) long. The shoots should be removed from the bushes and immediately prepared and inserted to half their length in the peat/sand mixture, either under mist or in a propagating case with a bottom heat temperature of 70°F (20°C). Strip the lower leaves from the cuttings, leaving three fully expanded leaves at the tips. When the cuttings have rooted, they should be potted on into an acid proprietary compost and hardened off, after which they should be grown in the open in the same way as other hardy potted plants.

Chapter 5 Cranberries

Introduction

Cranberry (*Vaccinium macrocarpon*), closely related to blueberry, is a creeping evergreen peat-loving prostrate shrub. The berries are ready for picking from late September in the south of England to mid-October in Scotland; they vary in size up to ½ in (1.2 cm) in diameter, are oval in shape and vary in colour from bright to dark red. The flesh is firm, dry and with a distinctly bitter taste. The fruit was once popular in this country when large quantities were imported from Russia and Scandinavia. Wild cranberries (*V. oxycoccus*) were once picked in northern England and southern Scotland and sold in local towns. Fruit which is now on sale in supermarkets and greengrocers is imported from North America.

Cranberries grow naturally in peat bogs which have acid soils and are well supplied with water; zones of hardiness 4-8. It is possible to imitate these conditions in the garden and once a raised or sunken bed is established it requires very little attention.

Varieties

At least one variety trial has been conducted in the United Kingdom and this demonstrated that North American varieties grow satisfactorily in the UK. The Scottish Crop Research Institute showed that 'Franklin' and 'CN' were outstanding and yielded between 1-1½ lb per sq yd (0.5-0.75 kg/m^2). An efficient nurseryman may not necessarily grow, but should be able to obtain, good quality plants of the required variety when a definite order is placed. If it is not possible to buy 'Franklin' or 'CN', there are other satisfactory but lower yielding varieties that are worth planting.

'Ben Lear'

An early-ripening variety that will only keep for two weeks at normal temperatures. The berries are large and dark red in colour.

'Bergman'

A mid-season variety of medium red colour. The yield is satisfactory.

'CN'

A large, bright red berry that crops very heavily.

'Franklin'

The berries are rather small and dark red in colour, but the yield of fruit is still very good. (See Plate 9.)

'McFarlin'

A late-season variety. The berries are large, dark red and round-oblong with a thick waxy bloom.

Growing Conditions

Cranberries must be provided with an acid growing medium that is retentive of moisture but at the same time is not waterlogged. As the natural rainfall is unlikely to keep the plants adequately supplied with water throughout the summer, it is essential that mains water free from lime should be available for application to the bed with a garden sprinkler. Applying water with a watering-can is impracticable and a hosepipe should not be used as it would disturb the sand and peat too much.

Bed Preparation

The most satisfactory width of bed is approximately 4 ft (1.2 m) so the plants can be tended without having to stand on the bed. It could be as narrow as 2 ft (0.6 m) to accommodate two rows of plants. The length of a bed is immaterial.

The soil on which the bed is to be sited or which is to be incorporated in the bed should be reduced to pH 4.5 in accordance with the instructions for preparing soil for growing blueberries (see p. 53). There is a choice of making a raised or sunken bed; the former is simpler to make, but the latter is more satisfactory as it conforms to the general level of the garden and is easier to water and maintain. However, sawdust should not be used for cranberries. To make a raised bed, it is then only necessary to place 6 in (15 cm) of horticultural peat, which contains no added nutrients or lime, on top of the area of treated soil that is to be planted with cranberries.

A sunken bed should either have a sub-soil that drains excess water from the overlying topsoil or, if there is any doubt about its ability to drain naturally, then artificial drainage should be provided. If the natural drainage is perfect, the bed should be excavated to a depth of 6-8 in (15-20 cm); if imperfect, to a depth of 10-12 in (25-30 cm). A tile or plastic drainpipe should be sunk into the floor of the bed to lead surplus water to a ditch or sump. The bed should then be covered with lime-free stones or brickbats over 1 in (2.5 cm) in diameter to a depth of 3 in (7.5 cm). These should be covered with a sheet of polythene mesh to prevent the compost or peat, with which the bed will be filled, from clogging up the drainage material. For this purpose old polythene compost bags in which numerous drainage holes have been made would be suitable. The bed should be filled with either horticultural peat, two parts of peat and one part of soil, or three parts of peat and one part of coarse sand. Heavy clay soils should not be used as they are almost impossible to mix satisfactorily with the peat.

Figure 5.1: Bed Preparation – Cranberries: (A) raised bed; (B) sunken bed

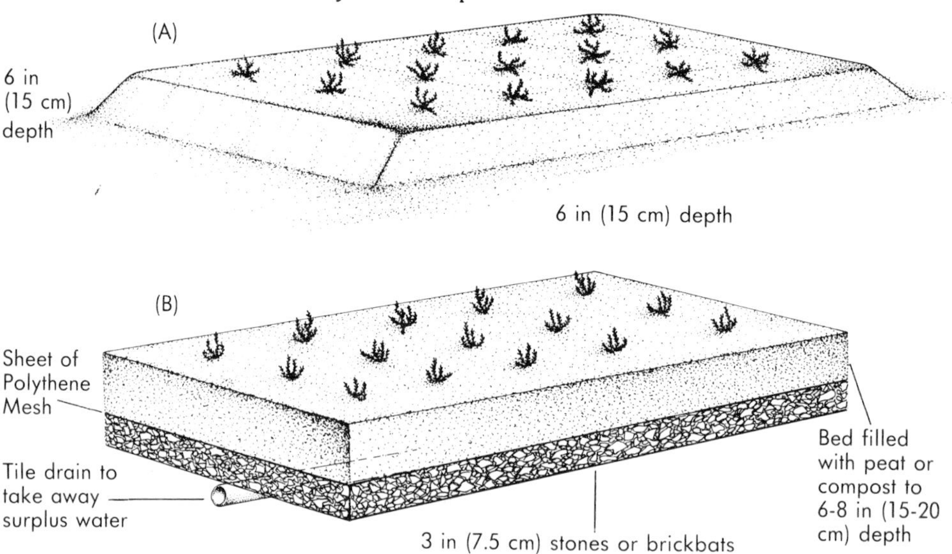

Planting

Plants growing in pots are likely to be the only type of stock available for purchase. These should be planted into the peat or compost at any time during the winter.

Propagation

Once initial stocks of varieties have been obtained, it should be possible to extend a bed or establish new beds with either unrooted or rooted cuttings.

In March, shoots 3-4 in (7.5-10 cm) long should be cut off the established plants and inserted to one-third of their length into the peat or compost. Two or three cuttings should be inserted together at each planting position.

To root cuttings, take shoots of the new season's growth when they are 2-3 in (5-7.5 cm) long in June or July (UK). They should be inserted into a compost of equal parts of sand and horticultural peat in a shaded frame. The compost in the frame should be damped down daily with water until the cuttings root, which they do very easily. Once rooted, they may be transplanted into the prepared bed or moved on into pots and grown on to provide large plants for planting in the bed.

The distances for planting cuttings or potted plants is 12 in (30 cm) each way.

Sanding

After planting, the bed should be covered with 1 in (2.5 cm) of coarse lime-free sand. This is not absolutely necessary but it prevents the surface of the peat from drying out, it gives some weed control, as fewer weeds germinate in sand than do in peat, and it provides a medium in which the cuttings root easily.

Watering

After planting the bed should be watered with a garden sprinkler until the peat is saturated. When a bed is established with cuttings, the sprinkler should be turned on five minutes each morning for the following 14 days.

Manuring

During the first year the objective should be to obtain rapid growth by the judicious application of fertiliser to get a good cover of the bed with new shoots. Harm could be done by exceeding the following rate of application. Broadcast over the bed:

April	⅔ oz sq yd (24 g/m^2)	Phostrogen
May	⅓ oz per sq yd (12 g/m^2)	Phostrogen
June	⅓ oz per sq yd (12 g/m^2)	sulphate of ammonia
July	⅓ oz per sq yd (12 g/m^2)	sulphate of ammonia

Wash off any fertiliser that has lodged on the plants with a short sprinkling of water.

Annual Culture

Vines are the horizontal shoots that creep over the surface and extend the area of the bed. Upright shoots grow from the vines. These are 2-3 in (5-7.5 cm) long, they flower during June (in the UK) and bear fruit in September/October (UK). Pruning should be carried out after harvest, cutting off any vines that have been pulled to the top of the bed and any upright shoots that have been damaged during picking. When the bed is fully established and covered with shoots, pruning carried out in March should consist of judiciously thinning out vines and upright shoots that are overcrowded and trimming the edges of the bed. Three to five years elapse before the plants fully cover the surface of the bed.

Sanding an Established Bed

Every four years a further ⅓ in (1 cm) depth of sand should be applied to the bed in early winter. This is not absolutely necessary but it can be beneficial by encouraging the vines and uprights to form more roots, and when necessary grow more vigorously.

Manuring an Established Bed

If, at the end of March (UK), the peat in the bed has not been saturated by winter rainfall, the bed should be watered until the peat is saturated. During April, broadcast over the bed:

⅔ oz per sq yd (24 g/m^2) Phostrogen

Watering an Established Bed

The amount of water that a bed requires depends upon the natural rainfall that occurs. Beds in gardens situated in the wetter northern and western districts of the United Kingdom will require less additional water than those situated in the drier districts of central and south-east England. It would be unlikely for beds made up with a large proportion of peat to require an application of water before the middle of June. If, after that, a period of two to three weeks drought occurs, the bed should be saturated with a garden sprinkler. It would be better to err on the side of over-watering provided that the bed had good drainage when it was first made.

Picking

In the earlier districts, the berries ripen from late September onwards. Although they can be picked over by hand when the first ripening berries develop their full colour, it is probably better to wait until all the crop is ready for picking as it is a back-breaking task, even worse than picking blackcurrants singly. In America, the berries are combed off the plants with a scoop-shaped container that has set on its lower lip a comb-like arrangement, the teeth of which are set wide enough apart to pull the berries off the shoots but to allow the shoots to pass through with a minimum of damage. It should be possible to make a similar gadget from a plastic dustpan or other plastic container of suitable size and thickness. The bed should always be combed in the same direction so that the stems are disturbed as little as possible.

Providing all damaged fruits are removed, cranberries will keep for two to three weeks at moderate room temperatures, for two to three months in a refrigerator at 35°-40°F (2-4°C) and almost indefinitely deep frozen. They require no preparation apart from being dry and being placed in tight-fitting polythene containers.

Pests and Diseases

Although cranberries are attacked by numerous pests and diseases in North America, none of these have so far been recorded in the United Kingdom. The only trouble that may have to be dealt with could be an occasional infestation by aphids.

Chapter 6 Gooseberries, Red and White Currants

Introduction

The gooseberry lost some of its popularity during the Second World War, when sugar was rationed, and it has never regained its appeal. Two other reasons for this, however, may be that the bushes must be sprayed to prevent mildew infection of the leaves and fruit and the leaves may be eaten by sawfly caterpillars. When picked hard and green gooseberries are an excellent fruit for freezing without any preparation; they make excellent jams and pies and when left on the bushes to become fully ripe, make a very good dessert fruit. They also mix well with raspberries in pies and tarts. On the other hand, redcurrants are the easiest fruit to grow and rarely require spraying with insecticide. These crops grow well throughout the United Kingdom, and in zones of hardiness 7-8.

Varieties of Gooseberry

At one time hundreds of varieties of gooseberry were grown in private gardens and on allotments, mainly in the Midlands and North of England, where gooseberry fruit competitions were keenly contested. Now the number of varieties listed in nurserymen's catalogues is restricted to two or three of each of the different coloured varieties. The Worcesterberry, mistakenly considered by some to be a hybrid of gooseberry and blackcurrant is, in fact, *Ribes divaricatum*. It should be grown in the same way as gooseberry.

Red Varieties

'Crown Bob'

A large oval, thin-skinned fruit that is slightly hairy and of good flavour. It ripens to a dark claret red colour but may be picked green. The bush has a spreading habit.

'Captivator'

The advantage of this variety is that it is almost thornless. It is believed to have originated in France and was introduced for the first time into The United Kingdom in 1981. The bush has a vigorous spreading habit and bears small berries, ¾ in (2 cm) in diameter. The berries are dark burgundy-red and sweet when fully ripe.

'Lancashire Lad'

The berries are medium large, oblong to oval in shape and hairy. Ripening mid-season, the flavour is fair, berries juicy and dark red. It is not suited to poor soils where growth may be stunted but on good soils crops heavily. It has some resistance to mildew.

'Whinhams Industry'

This variety, one of the best, is grown quite widely as it does well on most soils. Ripening mid-season it crops heavily and the berries are medium to large in size, oval in shape, hairy and possessing a good sweet flavour when ripe. Very susceptible to mildew.

White Varieties

'Careless'

'Careless' is the most widely grown variety as it is both reliable and heavy cropping. The berries are large, oval in shape, smooth-skinned and a green milky white colour when ripe. East Malling Research Station (EMRS) has selected a virus-free stock of 'Careless' and given it the name of 'Jubilee'. Preference should be given to growing this virus-free stock. An excellent fruit both for cooking and jam making, but the flavour of the ripe fruit is weak.

'Invicta' (EMRS)

'Invicta' is a new variety, bushes of which are now becoming available for purchase. In trials it has cropped twice as heavily as all other varieties, including 'Careless,' and is resistant to mildew. The growth is very vigorous, spreading and very thorny; the large oval berries are pale in colour, well flavoured and smooth-skinned. (See Plate 10.)

Yellow Varieties

'Golden Drop'

This variety is early-mid-season, has an upright compact habit and is easier to grow than 'Leveller'. The berries, medium to small in size, oval in shape, have a slightly downy skin. The colour is greenish-yellow and the flavour is good.

'Leveller'

'Leveller' has the reputation of having the best flavour of all gooseberries, but requires a good deep, well-drained soil. As it is difficult to grow, bushes should be reserved for the production of ripe berries. To produce large fruits the bushes should be hard pruned or the berries thinned early in the summer. The berries ripen mid-season, are oval in shape and the skin is smooth. The bush has a spreading habit and is susceptible to mildew.

Green Varieties

'Howards Lancer'

A mid-season to late regular heavy cropping variety that bears medium-sized roundish to oval-shaped berries. They are downy, thin skinned, pale green in colour and well flavoured. The bush grows vigorously with a spreading habit, produces a lot of suckers and is susceptible to mildew.

'Keepsake'

Keepsake is a mid-season ripening variety but the berries swell very early in the summer so it is one of the earliest varieties to be ready for picking green for cooking. The berries are oval in shape, whitish-green when ripe, slightly hairy and of good flavour. The bushes bear heavy crops but are very susceptible to mildew.

Worcesterberry

The Worcesterberry is a strong grower, very similar to a gooseberry, and can be formed into a shapely bush. The berries are a little over ½ in (1.25 cm) in diameter, oval in shape, purplish-black in colour and can be left for picking until as late as September. They make good jam and freeze well.

Varieties of Red and White Currant

'Jonkheer van Tets'

This is a new Dutch variety that ripens its fruit early and bears heavy crops towards the end of July. The berries are an attractive red colour and borne on long strigs.

'Redstart' (EMRS)

'Redstart' is a new late variety that ripens its fruit three weeks after 'Jonkheer van Tets'. The bushes are moderately vigorous, erect and very heavy yielding. The berries are of medium size, of good colour and have a strong acid flavour so that they make a very good jelly for eating with lamb and mutton. Bushes will not be generally available for purchase until after 1986.

'Stanza'

This is another new variety that is now on sale. It is not quite so late-ripening nor as heavy cropping as 'Redstart'. The short trusses bear dark-coloured, firm variable-sized berries.

'White Versailles'

The use of white currants would appear to be restricted to adding to fresh fruit salads, although Mrs Beeton has a recipe for white-currant jelly. 'White Versailles' is the variety usually listed in catalogues and grows satisfactorily.

Planting Stock

At the present time, only stocks of 'Invicta', 'Jubilee', and 'Redstart' are available with either a Special stock or 'A' certificate for freedom from virus and other diseases. Some of the other traditional varieties may become available with an 'H' certificate that will indicate that, although they do not originate from virus tested stocks, they should grow and crop satisfactorily. All the traditional varieties, even when they show no symptoms, must be considered to be infected with virus disease. In spite of this, they usually grow and crop satisfactorily, though it is considered that if they could be freed from virus they would crop more heavily. The life of a properly managed bush should be 20 years or more.

Types of Bush

Bushes may be purchased either with single stems (see Figure 6.1), which are to be preferred because they are easier to pick and manage, or as a stool bush, which has several branches arising from the same position at ground level.

All varieties are sold as two- or three-year-old bushes or cordons (see Figure 6.2). It is only rarely that one-year-old bushes are offered for sale. Good quality two-year-old bushes or three-year-old cordons are the best buys.

One-year-old Bush. The stem should be straight, at least 8 in (20 cm) long, with two branches 6 in (15 cm) long and a relatively small root system.

Two-year-old Bush. The stem should be straight, at least 8 in (20 cm) long, with three or more branches situated at the top of the stem. Each branch should be at least 12 in (30 cm) long. There should be a moderately-sized root system to match.

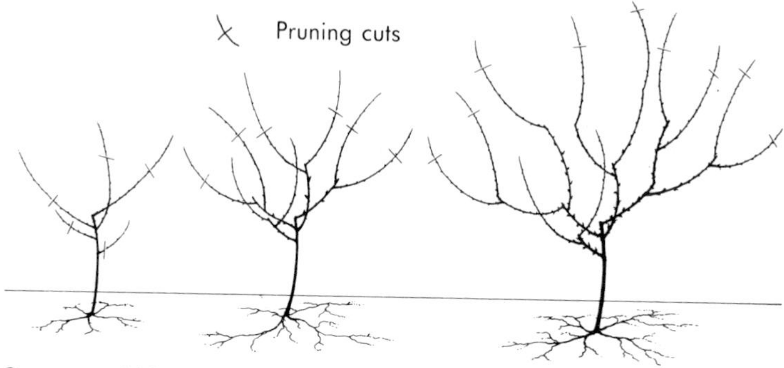

Figure 6.1: Examples of One-, Two- and Three-year-old Bushes, Showing Pruning Cuts for Gooseberry and Redcurrant

Three-year-old Bush. The bush should have an 8 in (20 cm) leg, at the top of which will be situated two or three main branches 3-4 in (7-10 cm) long, on which will be situated a small number of fruiting spurs. These main branches should be carrying five or more one-year-old branches, each 12 in (30 cm) long. The root system should be large.

Figure 6.2: Examples of Vertical and Inclined Cordons

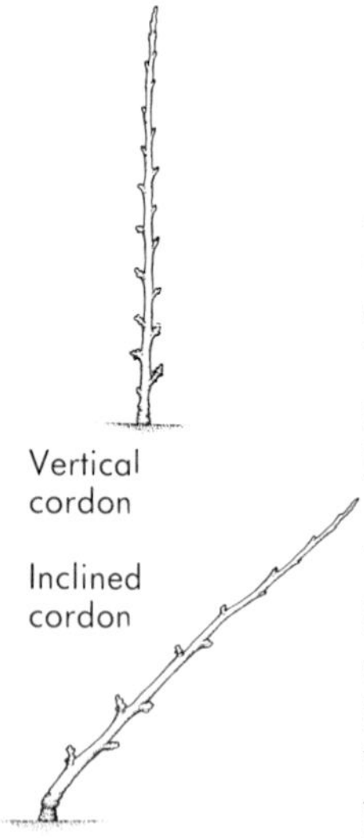

Vertical cordon

Inclined cordon

Single Cordons. A three-year-old cordon should have a single straight stem 30-36 in (75-90 cm) long that should be well furnished with fruiting spurs at approximate intervals of 3 in (7.5 cm) along its length. Before despatch by the nurseryman, the fruiting spurs are usually pruned but the leading shoot is left unpruned.

Double Cordons. A double cordon has a single stem 8 in (20 cm) long with two branches that grow out at right angles to the stem and opposite to each other for a distance of 6 in (15 cm). The two branches have then been trained vertically in the same form as a single cordon and each should be over 20 in (50 cm) long.

Treble Cordons. A treble cordon again has a stem 8 in (20 cm) long, and two side branches that grow out at right angles for a distance of 12 in (30 cm) before growing upright. A single branch also extends the main stem. The lengths of the three branches should be 20 in (50 cm) and their lower halves should be furnished with fruiting spurs.

Planting Distances

Bushes

Bushes should be plantd 5-6 ft (1.5-1.8 m) between the rows and 4-5 ft (1.2-1.5 m) between the bushes.

Cordons

Cordons should be planted against walls or fences and trained on wires or, in the open, trained on posts and wires. If more than one row, the distance between rows should be 4 ft (1.2 m).

Single cordons	12 in (30 cm)	Apart in the row
Double cordons	24 in (60 cm)	Apart in the row
Triple cordons	36 in (90 cm)	Apart in the row

Fans

A convenient, less time-consuming, but more rough-and-ready method of growing these fruits is as untrained fan-shaped bushes between pairs of wires secured to posts.

One- or two-year-old bushes should be planted 3 ft (90 cm) apart and 4 ft (1.2 m) between rows.

Planting

The stems of the plants (except those that are to be grown as stool bushes) should be carefully examined for the presence of dormant buds or white shoots. If these are not rubbed out, they later grow out into unwanted suckers (see Figure 6.3).

Figure 6.3: One-year-old Redcurrant Bush with a Good Root System and Possessing an Underground Bud that Should be Rubbed Off

Planting holes should be dug to a depth so that not more than one-third of the stem is below soil level after the soil has been firmed back on to the roots. If there are any roots on the stem above the soil, these should be cut off.

If the stool method of growing is adopted, the bushes should be planted deeply with the top of the stem at soil level.

Gooseberries, more than other bush fruits, benefit from planting in early winter and this helps to ensure that the shoots grow strongly during the following summer.

Weed Control

Weedex should be applied to the soil during February or after planting, whichever is the later (in the UK). In the years following, the Weedex should be applied each February, at a rate of application of:

Light soils – one sachet per gallon (4.5 l) applied to 50 sq yds (42 m^2)
Heavy soils – one sachet per gallon (4.5 l) applied to 33 sq yds (28 m^2)

If there are any established weeds or seedlings present, they should be killed by an application of one sachet of Weedol in 1 gallon (4.5 l) of water applied to 20 sq yds (17 m^2). Where there are no weeds present, there is no point in applying Weedol to clean ground. This herbicide should be applied with either a dribble bar or a coarse spray so that none falls on the gooseberry and redcurrant buds.

Manuring

It can be quite difficult to know just the right amount of nitrogenous fertiliser to apply to gooseberry and currant bushes, particularly in the three or four years after planting. Too little nitrogenous fertiliser applied and the bushes make no or insufficient new growth. On the other hand, if too much nitrogenous fertiliser is applied, the new growths become so long and soft that they are broken by the wind and other causes and the shape and size of the bush may be ruined as a result.

The maximum amount of fertiliser that should be applied during the March after planting in the UK and which should be applied in a circle 24 in (60 cm) diameter round each bush, should be:

⅓ oz (10 g) nitro-chalk
½ oz (18 g) sulphate of potash

or

1 oz (30 g) Growmore fertiliser

or

1 oz (30 g) Phostrogen

If, by the middle of June, the extension growth of the new shoots is not 9 in (22 cm) long, a further application of ⅓ oz (10 g) of nitro-chalk should be made.

In the year following in March, similar amounts should be broadcast over each sq yd (m^2), 3 ft (90 cm) either side of the bushes, provided that extension growths at the end of each year were approximately 18 in (45 cm) long. If they materially exceed this, the amount of nitrogenous fertiliser should be reduced substantially or omitted altogether. If the extension growths are substantially less than 18 in (45 cm), the nitrogenous fertiliser application should be increased.

When the bushes are fully grown, have filled the space allotted to them and are producing heavy crops, the rates of application should be increased to:

1 oz per sq yd (36 g/m^2) nitro-chalk
½ oz per sq yd (18 g/m^2) sulphate of potash
(Applied every fifth year:
2 oz per sq yd (70 g/m^2) superphosphate)

or

2 oz per sq yd (70 g/m^2) Growmore

or

2 oz per sq yd (70 g/m^2) Phostrogen

These rates should be adjusted according to the lengths of new growths of the lateral and leading shoots. The aim should be to maintain the length of these shoots at 9 in (22 cm).

Shaping and Training

Forming well-shaped, productive gooseberry and currant bushes or cordons can be quite a difficult task as every one is different from and does not conform to, instruction diagrams, which can only illustrate the principles of pruning. Also, the various varieties require differing treatments because each has a distinctive habit of growth.

Bushes

The primary objective should be to form an open vase-shaped bush with the branches equally spaced round the circumference. Sufficient space should be left between the branches to allow for easier picking of the fruit. Unless the bushes are grown in a cage, pruning should be left until the end of March (in the UK). By this time, birds should have stopped eating the buds and it should be possible to prune shoots back to sound buds.

First, any branch that cannot form part of the vase shape should be cut right out of the bush. Secondly, each remaining branch should be cut back to one-third of its original length and to right and left spacing buds. During the succeeding summer each pruned branch should give rise to two or three suitably spaced branches that can form part of the main framework of the bush. This method of pruning the branches should continue until the required 10-12 branches have been formed. In succeeding winters the leading shoot, which is the one at the end of each branch, should be pruned back to two-thirds of its original length until the bush occupies the space allocated to it. Shoots on varieties with an erect habit should be pruned to *downward* facing buds, whilst those on varieties with a pendulous habit should be pruned to *upward* facing buds.

The formation of a well-shaped bush should be hastened and assisted by judicious shoot thinning and summer pruning during June and July. All buds that grow out below the branches on the stem and appear from below soil level should be rubbed off or pulled up as soon as they appear. Shoots that grow strongly in the centre of a bush where they are not required or those that grow in competition with shoots that will form new branches should be cut right out in June or July when it becomes apparent that they are unnecessary for the development of the bush. Their removal improves the growth of the shoots that remain. The term 'summer pruning' involves cutting back all lateral shoots to encourage better fruit bud formation. It should be carried out towards the end of July or beginning at a time when growth is slowing down or has stopped. Each lateral shoot should be pruned back to one-third of its length. (See Figure 6.4.)

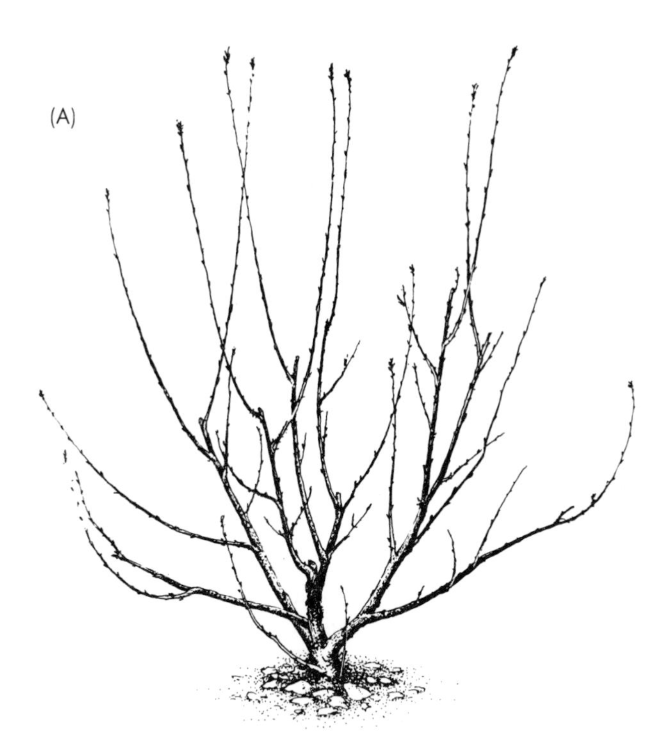

Figure 6.4: Three-year-old Redcurrant Bush: (A) before pruning; (B) after pruning

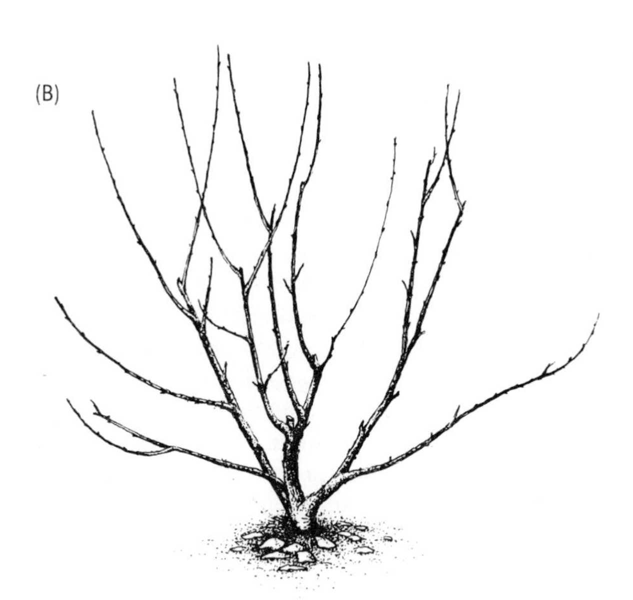

Winter pruning should be relatively simple where summer pruning has been correctly carried out. It consists of pruning back the leading shoots on the branches according to the age and habit of the bush and further pruning back each lateral to two or three buds.

Stooled Bushes

The pruning of gooseberries grown as stool bushes should be similar to that used for blackcurrants. However, given similar growing conditions, gooseberry shoots do not grow as long as those of blackcurrant, but two-year and older wood has more fruiting spurs on it than similarly aged blackcurrant wood. After planting, the shoots should be cut back to two buds. Provided that the soil is fertile and sufficient nitrogenous fertiliser has been applied, as many as twelve new shoots should appear and grow 18-24 in (45-60 cm) long. During the succeeding winter, a third of these shoots should be pruned back to the base, first removing any weak shoots, secondly any that droop too near the soil and thirdly thinning out the remainder to allow easier picking of the fruit. In each succeeding winter, a similar operation should be carried out, but first removing as much as possible of the shoots that bore fruit during the previous year, and leaving in the bush the maximum amount of new shoots either growing from the base of the bush or on short lengths of older wood. Shoots infected with mildew should be tipped back to healthy wood.

Single Cordons

Single cordons are best grown from one-year-old bushes or, failing a supply of these, from two-year-old bushes. The strongest shoot (which should be in direct line with the stem) should be selected and pruned back to half its original length. All other shoots should be cut back to a length of 1 in (2.5 cm). A bamboo cane should be inserted alongside the stem of the bush. During the summer, the new shoot that grows from the topmost bud should be trained and tied alongside the cane, several times, as it grows. The lateral shoots should be summer pruned during late July or August by cutting them back to one-third of their length. During June and July, it will become apparent that very strong growing lateral shoots are appearing on the cordons and that if left they could grow 2 ft (60 cm) long. If these are wrongly positioned, they should be cut or rubbed clean out. The objective should be to have a fruiting spur situated every 3 or 4 in (7.5 or 10 cm) along the stem of the cordon. Each winter the leading shoot should be cut back to half its length and the lateral shoots to 1 in (2.5 cm). When the cordon is fully grown, the leading shoot should be cut back to ½ in (1.25 cm).

Double Cordons

To train double cordons, two bamboo canes should be inserted in the soil each 6 in (15 cm) away from the stem of the newly planted bush. A second cane should be tied horizontally to these canes at a height of 8 in (20 cm) above the soil. The two strongest shoots should be selected and tied down to the horizontal bamboo in either direction. They should be shortened to half their length and all other shoots should be shortened to 1 in (2.5 cm). During the following summer, each leading shoot should be tied and trained along the horizontal bamboo and up one of the vertical bamboos. After this, each shoot should be treated as if it were a single cordon.

Treble Cordons

Three bamboo canes should be inserted in the soil, one next to the stem of the newly planted bush and the other two, 12 in (30 cm) on either side, but in line with the first bamboo. A fourth bamboo should be tied horizontally at a height of 8 in (20 cm) above the soil to the other bamboos. The three strongest shoots should be selected and pruned back to half their lengths; one should be trained up to the centre bamboo and the others should be trained either way along the horizontal bamboo. The method for training single and double cordons should then be applied.

Sucker Control

Gooseberry and currant bushes go on producing suckers throughout their growing lives. These suckers should be pulled off during June and July (in the UK) whilst they are still soft and weakly attached to the bush. They should not be cut off, as this leads to the production of an even larger number of unwanted shoots.

Watering

Gooseberries and red and white currants should be watered in times of drought in May and June but sparingly in July, otherwise the new shoots will become very soft and liable to be broken by the wind or during careless picking.

Picking

Gooseberries should be picked over a period of several weeks, starting as early as the end of May, when the berries are twice the size of peas and finishing as late as August, when the berries are very large and ripe. The objective should be to thin out the berries progressively, picking the larger berries on each occasion so that at the end of June there should be one large berry remaining at intervals of 2 in (5 cm) along each branch. These should be allowed to remain until they are fully grown and just begun to ripen. Then half should be removed for freezing, bottling and making into jam. The remaining half should be picked later, when fully ripe, for eating as dessert fruit. The berries can hang on the bushes for a remarkable length of time without becoming over-ripe or splitting.

Red and white currants, depending upon the variety being grown, ripen their berries during the period late July to mid August. They are usually cleared from the bushes in one picking, although there is no reason why the first ripening berries should not be picked singly as they ripen.

Pests and Diseases

American gooseberry mildew is a disease that can be expected to infect gooseberry bushes every year and against which control measures should be taken. It may also be necessary to spray with insecticide to kill aphids before flowering and sawfly caterpillars after flowering.

The only serious trouble on red and white currants is aphids.

Propagation

Red and white currants root very easily whilst gooseberries are usually very difficult and the percentage of cuttings that can be expected to root can be 50 per cent or lower. In order to be successful with gooseberries, it is very important that the cuttings should be taken in October or early November *before* all the leaves have fallen off the bushes. Strong, straight shoots 15-18 in (35-40 cm) long, that grew during the preceding summer, should be selected. The top few inches (cms) of unripened wood should be cut back to give a cutting 12 in (30 cm) in length. A firm planting bed should be prepared and the cuttings pushed vertically down into the soil to three-quarters of their length, leaving three buds exposed. On heavy soils, take out a narrow trench 9 in (22 cm) deep in the bottom of which a 2 in (5 cm) depth of coarse sand should be placed. The bases of the cuttings should be positioned in the sand and the soil replaced and firmed. The distance between the cuttings should be 6 in (15 cm) and the distance between rows should be 24 in (60 cm). The cuttings can be loosened in the soil by hard frost and should be re-firmed at the end of the winter. If space is not available in

October, the cuttings could be stored in bundles in a sandpit until as late as December.

When the cuttings begin to grow in late April or early May, they should be given a side dressing of nitro-chalk at the rate of ½ oz per sq yd (18 g/m^2). They should be sprayed at the same time as fruiting bushes to control aphis, mildew and leaf spot so that growth is not impaired.

Table 6.1: Gooseberry, and Red and White Currant Protection Chart

Problem	Chemical	Key	Product	April	May	June	July	Aug.	Sept.	Oct.
American Gooseberry Mildew	Bupirimate and Triforine	3	Nimrod T							
	Sulphur	5	Comac Cutonic Sulphur Flowable							
Aphids	Dimethoate	1	Boots Greenfly & Blackfly Killer							
		6	Murphy Systemic Insecticide							
	Fenitrothion	7	PBI Fenitrothion							
	Malathion	7	Malathion Greenfly Killer							
		6	Murphy Liquid Malathion							
	Permethrin	7	Bio Sprayday							
		2	Fisons Whitefly & Caterpillar Killer							
		3	Picket							
	Pirimiphos-methyl	3	Sybol 2							
Botrytis Grey Mould	Benomyl	3	Benlate plus Activex							
	Carbendazim	1	Boots Systemic Fungicide							
	Thiophanate-methyl	4	Fungus Fighter							
		6	Murphy Systemic Fungicide							
Gooseberry Sawfly	Fenitrothion	7	PBI Fenitrothion							
	Malathion	7	Malathion Greenfly Killer							
		6	Murphy Liquid Malathion							
	Rotenone	7	Liquid Derris							
Green Capsid Bug	Fenitrothion	7	PBI Fenitrothion							
		7	Malathion Greenfly Killer							
	Malathion	6	Murphy Liquid Malathion							

Pest/Disease	Active ingredient	Maker	Product
	Permethrin	7	Bio Sprayday
		2	Fisons Whitefly & Caterpillar Killer
		3	Picket
	Pirimiphos-methyl	3	Sybol 2
Leaf Spot	Copper	5	Comac Bordeaux Plus
		6	Murphy Liquid Copper Fungicide
	Benomyl	3	Benlate plus Activex
	Carbendazim	1	Boots Systemic Fungicide
Red Spider Mite	Malathion	7	Malathion Greenfly Killer
		6	Murphy Liquid Malathion
	Pirimiphos-methyl	3	Sybol 2
	Dimethoate	1	Boots Greenfly & Blackfly Killer
		6	Murphy's Systemic Insecticide
Winter Moth	Fenitrothion	7	PBI Fenitrothion
	Permethrin	7	Bio Sprayday
		2	Fison's Whitefly & Caterpillar Killer
		3	Picket
	Pirimiphos-methyl	3	Sybol 2

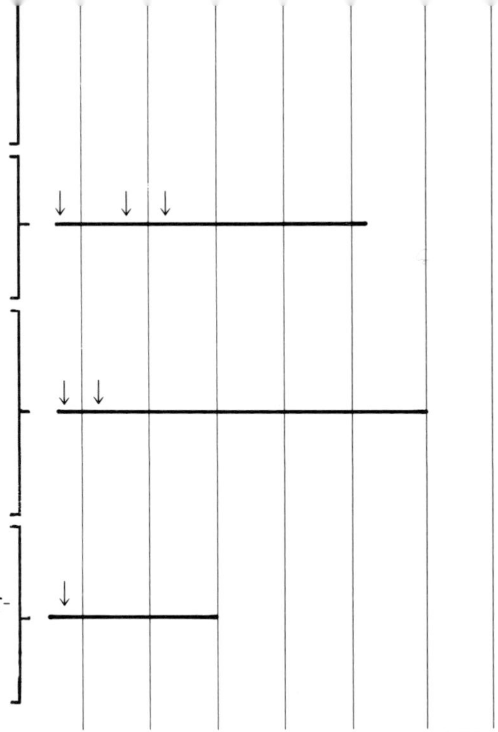

Key: —— Infestation periods ↓ = Timing of pesticide application

1 The Boots Company PLC, Nottingham.
2 Fisons PLC, Paper Mill Lane, Ipswich.
3 ICI Products, Woolmead House, Farnham, Surrey.
4 May & Baker, Regent House, Hubert Road, Brentwood, Essex.
5 McKecknie Chemicals Ltd, PO Box 4, Widnes, Cheshire.
6 Murphy Chemical Co., Latchmore Court, Brand Street, Hitchen, Herts.
7 Pan Britannica Industries, Waltham Cross, Herts.

Chapter 7 Kiwifruit

Introduction

The Chinese gooseberry (*Actinidia chinensis*) was introduced into New Zealand at the beginning of this century but it was not until the 1930s that its potential as an horticultural fruit was developed under its more familiar name of kiwifruit. It requires a mediterranean-type climate, with mild winters and warm or hot summers. Kiwifruits are hardy out of doors in most of zone 8 and all of zone 9. The kiwifruit can only be marginally successful in England and then in the most favourable situations. It will grow out of doors wherever grapes grow but during the winter the fruiting wood can be killed when temperatures fall to 14°F (−10°C), the amount of damage depending upon how well the wood had ripened off in the autumn. The newly emerged shoots and flowers will be severely damaged when night temperatures in spring fall to 29°F (1.5°C) for as little as an hour. At the Royal Horticultural Society's gardens at Wisley, in a favourable situation, frost kills the crop one year in four. The bush requires very good shelter from the prevailing winds, otherwise tender young shoots will be broken off and damage at any time to the leaves reduces the yield. The fruits are smaller, are less well flavoured and have a shorter storage period than imported fruits.

Varieties

The kiwifruit has the appearance of a miniature coconut and when grown under optimum conditions measures 3-4 in (7.5-10 cm) long and 1½ in (4 cm) broad. It has a brown furry coat but when cut open has a juicy, fresh green-coloured flesh with small brown seeds.

'Abbott'

An early-flowering variety that produces heavy crops. The fruit is medium-sized and covered with long dense hairs.

'Bruno'

Flowers slightly earlier than 'Abbott'. The fruit is fairly large,

1 Manganese Deficiency on Raspberry

2 Iron Deficiency on Raspberry

3 Potash Deficiency on Strawberry

4 Tayberry (photo: SCRI)

5 Tayberry Trained on Wall

6 Tummelberry (photo: SCRI)

7 Blackcurrant: 'Ben Sarek' Variety (photo: SCRI)

8 Blueberry: 'Bluecrop' Variety (photo: SCRI)

9 Cranberry: 'Franklin' Variety (photo: SCRI)

10 Gooseberry: 'Invicta' Variety

11 Raspberry Leaf Spot Virus on 'Glen Clova' Variety

cylindrical and darker brown than other varieties. It is a heavy cropper.

'Hayward'

'Hayward' would probably be the best variety to grow in the United Kingdom as it occupies over nine-tenths of the area grown in New Zealand, which would indicate that it is adapted to the widest range of growing conditions. Other characteristics that make it more suitable to the English climate are very late flowering, large fruits, superior flavour and good keeping qualities. The fruits are pale greenish-brown, covered with fairly fine silky hairs.

'Matua'

The kiwifruit is self-sterile, therefore it is necessary to plant a male plant to provide pollen for the female plant(s). 'Matua' would be the best male variety to plant. The ratio of male to female plants should be a minimum of one to seven.

Soil Conditions

Kiwifruits require deep free-draining soils that are retentive of moisture or, if not, can be watered when necessary. Poorly-drained, heavy clay soils and shallow or very light soils should be avoided. Soil pH should be 6.5 and the requirements for nitrogen, phosphorus and potassium are similar to those for blackberries (see below, under 'Manuring').

Climatic Conditions

The area of southern England, where kiwifruit could crop satisfactorily in the open, would appear to be south of the line drawn from London to Bristol. Against a south-facing wall, the area where they can be grown is south of the line drawn from King's Lynn to Shrewsbury – in both cases, in locations that are not affected by the cooling influences of the North Sea and not exposed to winds blowing off the English Channel. Within this area, kiwifruits should not be planted in frost holes or at the bottoms of river valleys where the flowers and new growth would be susceptible to damage by late spring frosts. A suitable garden should have a southerly or south-westerly aspect and maximum shelter from the prevailing winds.

Planting Distances

The kiwifruit plant is a very vigorous climbing vine. 'Hayward', the weakest grower, should be planted 18 ft (5.5 m) apart whilst all other varieties should be planted 22 ft (7.0 m) apart. The distance between two rows should be 10 ft (3.0 m). One male pollinating variety should be planted with seven female plants. As the number of kiwifruits that are likely to be planted in one garden is very small (less than four), one male pollinating vine takes up a disproportionate amount of productive space; therefore, where less than four vines are to be planted, one male and one female plant should be planted together in the same planting hole and their vines trained in opposite directions along the wires.

Planting

Kiwifruits are invariably supplied in pots and they should be purchased and planted in late March or early April when the risk of damage by severe winter frosts will have passed. A hole 2 ft (60 cm) square should be dug to a spade's depth and ½ oz (15 g) of Growmore or Phostrogen fertiliser mixed with the soil. The plant(s) should be planted in the centre of the hole and the soil well firmed. The planting position should be midway between posts.

Manuring

Kiwifruits require generous applications of manure if they are to grow the long length of vines that are required to furnish the wirework.

During the March following planting, broadcast on the soil in a circle 18 in (45 cm) diameter round each plant:

1 oz (30 g) nitro-chalk
⅓ oz (10 g) sulphate of potash

or

2 oz (60 g) Phostrogen or Growmore

Similarly, at the end of May and June, broadcast:

⅔ oz (20 g) nitro-chalk

In the succeeding years during March, broadcast on each sq yd (m^2) to a distance of 3 ft (90 cm), on both sides of the rows:

⅔ oz (24 g) nitro-chalk
⅓ oz (12 g) sulphate of potash

If the new vines do not grow to a satisfactory length, increase the amounts of nitrogenous fertiliser applied by 50 per cent. If the growth is too vigorous, reduce the amounts applied by 50 per cent.

Training Systems

Kiwifruits can be grown on wires against a wall or fence or on posts and wires. Such a support system should be similar to that required by blackberries but the wires should be placed 18 in (45 cm) apart. Vines trained in this way have to be spur pruned. Alternatively, vines may be trained on a pergola or on overhead wires supported on 'T' posts (see Figure 7.1). These enable the New Zealand method of pruning to be adopted in which the fruits are borne on one-year-old wood.

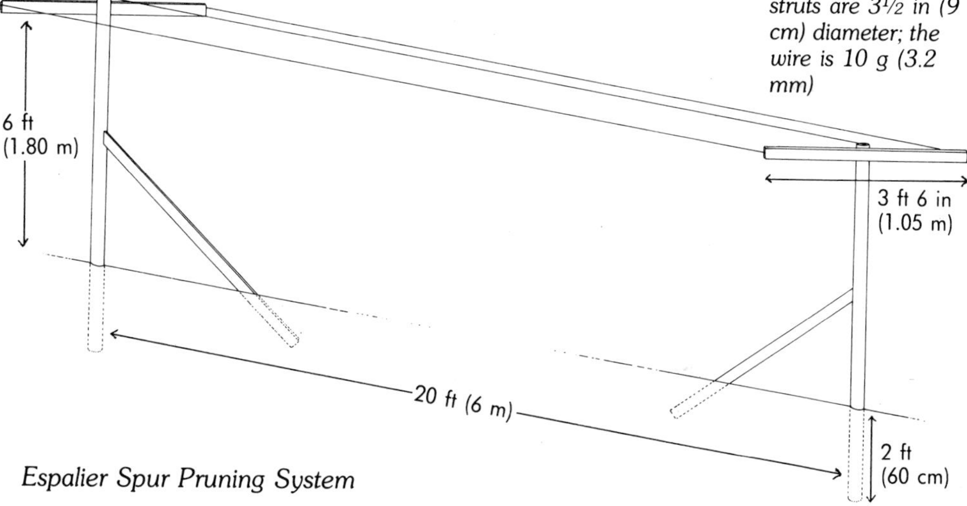

Figure 7.1: 'T'-bar Framework for Growing Kiwifruit. The uprights are 4½ in (11 cm) diameter; the cross pieces are 4 x 2 in (10 x 5 cm); the struts are 3½ in (9 cm) diameter; the wire is 10 g (3.2 mm)

Espalier Spur Pruning System

A 7 ft (1.8 m) long bamboo cane should be pushed into the soil next to the vine and tied to the wires. The vine should be cut back at the height of the lowest wire. The three strongest shoots that grow should be selected and any others pinched out. The upper one, as it grows, should be trained and tied to the bamboo cane but should not be allowed to twist itself round the cane or its growth could be restricted. The remaining shoots should be trained in opposite directions along the wire. The procedure should be repeated as the centre vine reaches each wire.

The laterals should have their growing points pinched out when they have grown 3 ft (1 m) long. This encourages the production of side shoots and these should be pinched back to five leaves to form fruiting spurs; subsequent sub-laterals should be removed as they appear. This procedure encourages the formation of fruit buds. When spurs bear fruit, vegetative shoots should be regularly pinched back to seven leaves. During winter, cut back fruiting spurs to two buds beyond where the fruit was borne in the preceding season (see Figure 7.2).

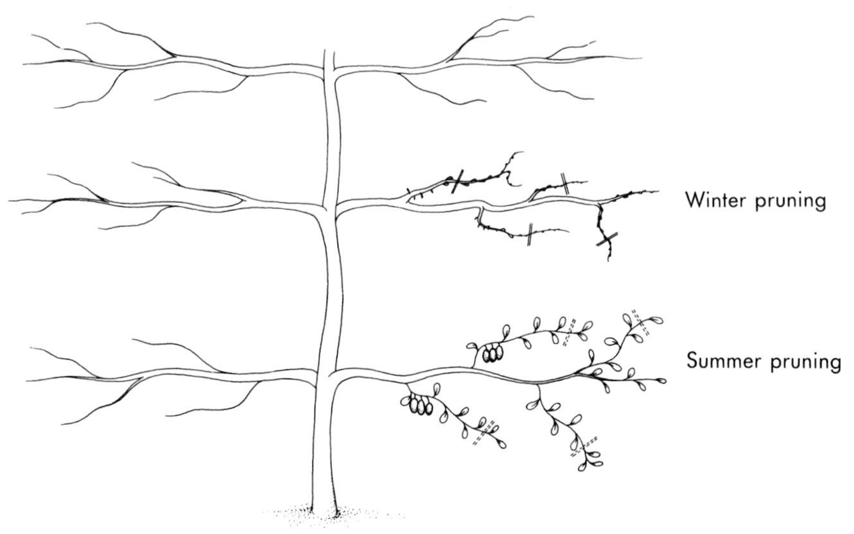

Figure 7.2: Espalier Spur Pruning System for Growing Kiwifruit

The New Zealand Renewal System

A bamboo cane 7 ft (2.1 m) long should be pushed into the soil next to the vine and secured to the centre wire. The vine should be cut back to a height of 18 in (45 cm). The strongest shoot that grows should be selected and trained up but not allowed to twist round the bamboo cane. The growing point should be pinched out when it reaches the wire. The two strongest growing shoots that appear should be selected and trained in opposite directions along the wire but should not be allowed to twist round the wire. These are known as the leaders. Lateral growths should be thinned out so that a pair of laterals or fruiting arms are situated along the leaders at intervals of 20 in (50 cm) (see Figure 7.3). The fruiting arms should be trained horizontally and opposite to each other to the outer wires and their growing points pinched out. The building of the basic framework continues in this manner until the wires or pergola are furnished. Where a male and female are grown together, the main trunk of each is trained up to the top wire where one leader of the male is trained in one direction and one leader of the female in the other direction.

Fruiting laterals that grow on the fruiting arms grow downwards and, in the year following, flower and bear fruit. The objective of summer pruning is to prevent the vines from becoming

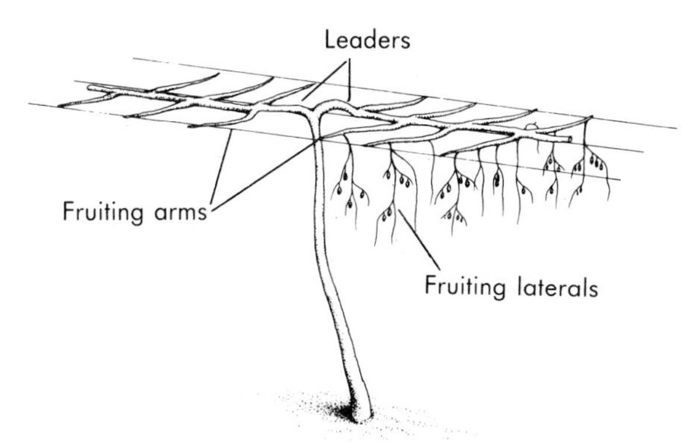

Figure 7.3: New Zealand Renewal System for Growing Kiwifruit

overcrowded and to select suitably placed new shoots that will bear fruit in the succeeding year. The shoots that extend the fruiting laterals should be pinched back to six leaves. Any erect water shoots should be cut back to a stub 1 in (2.5 cm) long. One new lateral shoot at or near the fruiting lateral should be selected and allowed to grow unchecked, in order to bear fruit in the following year. Summer pruning should be carried out several times during the summer to maintain this basic structure, surplus intertwining shoots should be cut out and all shoots that grow down below knee height should have their growing points pinched out.

Winter Pruning

Winter pruning should be left until the end of the winter so that any shoots that have been killed by frost can be removed. Any fruit spurs should be pruned back to two buds beyond where the last fruits were borne. The arms that bore fruit in the previous year should be cut back to their replacement shoots. Approximately three fruiting laterals with plump, closely-spaced buds should remain on each fruiting arm to bear fruit. Those in excess of this number, and that have flat widely-spaced buds, should be cut out.

Manuring

Kiwifruits can be slow to grow away in the first summer after planting. In April, May and June, top dressings of ½oz (15 g) of nitro-chalk should be broadcast round each plant in a circle of 3 ft (1 m) diameter. In the years following, broadcast 4 ft (1.2 m) either side of the row, 1 oz per sq yd (36 g/m^2) Growmore or Phostrogen.

Weed Control

The soil in which kiwifruits grow should be kept free from weeds, either by shallow hoeing or by the applications of herbicides.

Figure 7.4: Kiwifruit: Winter and Summer Pruning Cuts

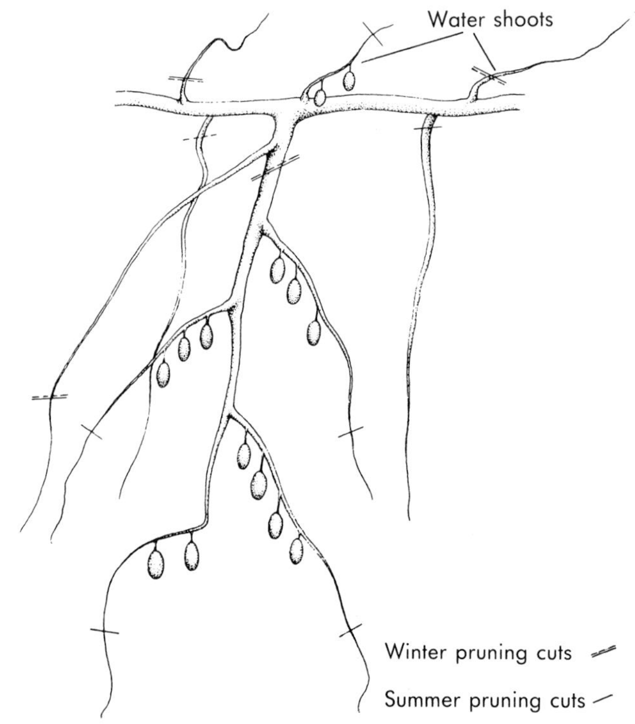

Weedex should be applied to the soil in February – see page 19. If Weedol is applied to kill established weed seedlings, care should be taken to ensure that none falls on the stems of the vines: these are able to absorb this herbicide and incur severe damage.

The rate of application of Weedex is:

Light soils – one sachet per gallon (4.5 l) applied to 50 sq yds ($42\ m^2$)
Heavy soils – one sachet per gallon (4.5 l) applied to 33 sq yds ($28\ m^2$)

Pollination

Although bees and other insects will transfer pollen from the male to female flowers, hand pollination ensures a good set of fruits and that they individually swell to maximum size. Male flowers that have just opened contain the largest number of pollen grains and female flowers remain receptive for up to eight days after opening. When both types of flowers are dry, whilst holding the female with one hand, gently brush (with a circular motion) the anthers of the male flower over the stigmas with the other hand. One male flower should be used to pollinate five females. Pollination should be carried out on three occasions, when 30 per cent, 50 per cent and 100 per cent of the flowers are open. More frequent pollination than this can have an adverse effect on the fertilisation of the flowers.

Watering

The supply of water should be maintained in times of drought from May to July but should, if necessary, be applied sparingly in August and September. It is important that the vines should ripen slowly and be very mature before low winter temperatures occur.

Picking

Kiwifruit are easily harvested by snapping the stalks which remain on the vine. They will be ready for picking when it is first possible to make a slight depression on the surface when pressing hard with the thumb. The vines can be picked over three or four times, picking the larger fruits on each occasion, leaving the smaller ones to sweli. When early autumn frosts occur and may damage the fruit, the vines should be cleared at one go. Stored at room temperature they will keep for 10-14 days and in a refrigerator for two to three months. Stored fruit should be kept away from any other stored fruit, particularly apples, as the production of ethylene by these fruits will rapidly lead to the ripening of kiwifruits. On the other hand, when a bowl of ripe kiwifruit is required, a ripe apple should be placed in the bowl with them.

Pests and Diseases

The only pests that are likely to infest kiwifruit in the United Kingdom are the general fruit pests; aphids, capsid bug, caterpillars and red spider mites. So far, none of the pests or diseases specific to kiwifruits have been recorded.

Chapter 8 Raspberries

Introduction

Raspberries can be grown successfully almost anywhere in the United Kingdom, provided they are sited on satisfactory soil with good drainage. They will commence to crop in their second growing season following planting and, because they flower later than most other fruits, they tend to escape damage by spring frosts. To obtain worthwhile crops of raspberries it is necessary to buy good quality Ministry Certified canes for planting, provide a framework of posts and wires on which to support the canes and be prepared to spray the canes during the summer for pests and diseases. As long as raspberries are kept free of mosaic virus, they can be expected to last for nine to ten years without needing to be replaced.

If raspberries are neglected, they quickly become a forest of weeds intermingled with new and dead canes and, unless they are sprayed at least once annually to control the larvae of the raspberry beetle, the fruit will become affected and spoiled by this pest. They grow best in zones of hardiness 5-8.

Summer Fruiting Varieties

In recent years there have been a number of new varieties introduced and more may be expected from the breeders at the East Malling Research Station (EMRS) and the Scottish Crop Research Institute (SCRI). These are rapidly replacing all the old traditional varieties and never has there been such a wide choice of so many good quality heavy cropping raspberry varieties. The following descriptions will enable readers to choose varieties to suit their particular needs. The seasons of ripening are shown in Table 8.1. The varieties marked with an asterisk (*) are resistant to colonisation by aphids.

*'Delight' (EMRS)**

'Delight' is the largest fruited and one of the heaviest cropping varieties ever introduced. The fruits are very large, blunt-conical in shape, pale red in colour with a thin skin that breaks and bleeds very easily. It requires frequent picking. Opinions differ about the flavour that at best can only be described as moderate. The fruits are unsatisfactory for bottling, jam making and freezing.

'Delight' is a very vigorous variety with tall, very thick canes. Properly managed, it will produce a heavy crop in the second year after planting. Surplus canes should be rigorously suppressed each spring and summer.

'Glen Clova' (SCRI)

'Glen Clova' was the first of the newer selections to be named because it was early in ripening and heavy cropping. The berries are medium-sized, round-conical in shape but they are very numerous because two fruiting laterals are produced where other varieties produce only one. The colour of the fruit is bright red with an acid flavour. It is good for bottling, jam making and freezing but requires plenty of sugar when eaten fresh.

This variety is tall growing and produces an over-abundance of new canes. On fertile soils it is necessary to control this vigour by removing the first flush of canes (see page 100) if the ripe fruits are not to be hidden from view by the new canes and the full yield is to be realised.

For the amateur, 'Glen Clova' has two serious disadvantages. It is susceptible to infection by cane midge blight that gives rise to dead or wilting fruiting canes during the spring and summer. It is difficult for the amateur gardener to control the pest that gives rise to this trouble. 'Glen Clova' is also susceptible to infection by raspberry leaf spot virus (see Plate 11). This virus causes a yellow spotting and distortion of the leaves followed by death of the bush. Raspberry leaf spot virus is carried and tolerated by many raspberry varieties including 'Malling Jewel' in particular. It is inadvisable to plant 'Glen Clova' in the neighbourhood of 'Malling Jewel' and many other varieties. It is safer to grow it on its own.

*'Glen Moy' (SCRI)**

It will not be possible to buy 'Glen Moy' and 'Glen Prosen' canes until 1985. 'Glen Moy' is an early variety that yields as heavily as 'Glen Clova' but has a more concentrated ripening period. The berries are medium-sized, medium red in colour and short-conical in shape. It has a good flavour and is suitable for all purposes. The canes are moderately vigorous, stout and upright. The berries are borne on short stout laterals. The new canes are smooth and spineless but as the bark splits, they could be affected by cane midge blight. 'Glen Moy' is susceptible to leaf

Table 8.1: Ripening Periods of Summer Fruiting Raspberries

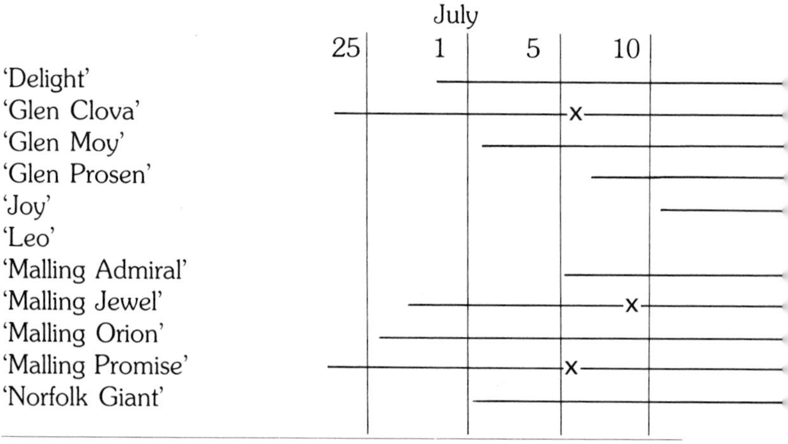

	June 25	July 1	5	10
'Delight'				
'Glen Clova'				x
'Glen Moy'				
'Glen Prosen'				
'Joy'				
'Leo'				
'Malling Admiral'				
'Malling Jewel'				x
'Malling Orion'				
'Malling Promise'				x
'Norfolk Giant'				

Note:——x—— = 50 per cent of the fruit picked.

spot virus but its resistance to aphid colonisation could prevent this disease becoming a problem. The buds at the tips of the new canes can come into flower in the autumn and in the south of England bear ripe fruits. (See Plate 12.)

*Glen Prosen (SCRI)**

In many respects 'Glen Prosen' is similar to 'Glen Moy' and it can be difficult to distinguish between the new canes of the two. It is a mid to late season variety with a long picking period. The round-conical berries are medium red in colour with good flavour and of exceptional firmness. The canes are spine free and laterals are stout and medium long. 'Glen Prosen' is resistant to aphids, and infection by leaf spot virus, if it occurs, does not give trouble.

*'Joy' (EMRS)**

'Joy' is a newly released variety that has just become available. The canes are tall, spiny and moderate in number. The very long strong laterals are held horizontally and covered with very sharp spines. 'Joy' yields more heavily than 'Leo' but not as heavily as 'Glen Clova', bearing large, conical deep red berries that are of good flavour. This variety will bear heavier crops if double planted, i.e. by putting two canes together in each planting position.

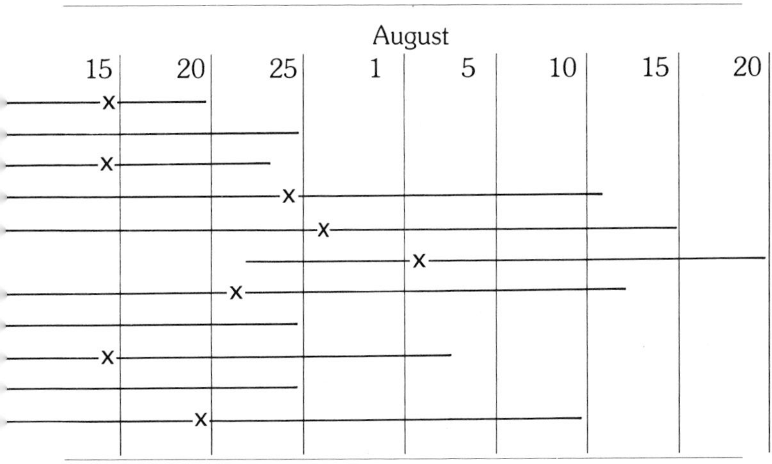

August

15 20 25 1 5 10 15 20

*'Leo' (EMRS)**

This is a new variety that has been available for four years. It is even later ripening than 'Joy' and replaces the old variety 'Norfolk Giant'. 'Leo' produces a moderate number of tall canes and heavier early cropping can be induced by double planting. The canes are moderately spiny and the long laterals droop in the rows. The berries are large, roundish, medium red in colour, firm and with a good aromatic flavour. The yield is not high but is greater than that of 'Malling Jewel'.

'Malling Admiral' (EMRS)

This is an outstanding variety for fruit quality but its yield is low, similar to that of 'Malling Jewel'. The berries are large, conical in shape, dark red in colour and have an excellent flavour. The fruit when frozen, bottled or made into jam provides a first-class product.

'Admiral' has numerous tall strong canes and these produce long laterals that are liable to be broken by the wind in exposed situations. The canes invariably remain dark brown in colour and are resistant to infection by diseases. If only one variety is to be grown, 'Admiral' is the one that should be selected.

'Malling Jewel' (EMRS)

'Malling Jewel' has been the standard variety for 20 years because of its longevity due to tolerance to virus infection and consistent though moderate yield of fruit. The flavour of the fruit, though not outstanding, is adequate. The firm, dark red conical berries hang for over a week on the bushes without becoming too ripe. The yield is only moderate and the fruit ripens over a short period of three to four weeks. To obtain high yields in the years immediately after planting, two canes should be placed in each planting position. The bushes are easy to manage as only a moderate number of canes grow each year. The stiff, stout laterals display the fruit beyond the neat foliage for easy picking.

*'Malling Orion' (EMRS)**

Although 'Orion' has the potential to crop very heavily it seldom does because the excessive number of canes that it produces are very susceptible to infection by diseases. The fruits are an attractive bright red colour but small in size, 'Orion' has been superseded by better varieties.

'Malling Promise' (EMRS)

'Promise' was grown extensively from 1950 until 1960 and is still included in some nurserymen's catalogues. It can produce heavy crops of large, bright red but soft, crumbly berries early in the season. However, the poor fruit quality and problems with disease have enabled newer varieties to supersede it.

'Norfolk Giant'

Cultivated since 1926, 'Norfolk Giant' was and is well known as a good quality but low yielding, late-ripening variety. It is being superseded by 'Leo' because of its small berry size, poor yield and susceptibility to raspberry ring spot virus, which is soil-borne and causes the death of the canes or bushes.

Autumn Fruiting Varieties

*'Autumn Bliss' (EMRS)**

The breeding of this new variety, which is expected to be available for purchase in 1987, is an advance as far as autumn fruiting kinds are concerned. The berries commence ripening in mid-August and continue doing so until the end of September. This early period of ripening enables 4 lb of fruit to be picked from each yard of row (2 kg/m) which is three times greater than the yield from the later-ripening, currently available varieties. This yield is only one-third lower than that given by a good summer fruiting variety. The berries are an attractive medium red colour, of good size and satisfactory flavour and firmness. The canes are short and sturdy and may not need supporting.

'Fallgold' (USA)

This is a yellow fruited variety with a sweet mild flavour and as it ripens from September onwards, cannot be heavy yielding. The canes are moderately vigorous and numerous.

Heritage' (USA)

As the berries ripen from mid-September and the canes are very vigorous, this variety requires good shelter and good support from the prevailing winds. The yield is poor as only half the crop ripens before temperatures become too low which makes it unsuitable for northern districts. The fruit is of medium size, very firm and of moderate flavour.

'September' (USA)

Although 'September' begins to ripen its berries in mid-August, yields have been disappointing, probably because stocks have been infected with virus. The canes are very vigorous and require support. The berries are medium-sized mid-red colour and of moderate flavour. They do not plug very easily.

'Zeva Herbsternte' (Switzerland)

This variety, usually known as 'Zeva', is of medium vigour and produces a multitude of canes, the spread of which should be rigidly restricted. The berries, which are somewhat hidden by the leaves, are large, medium to dark red in colour and of good flavour. They are inclined to break up if not pulled carefully off the plug when fully ripe.

Cultivation of Summer Fruiting Varieties

Soil

Raspberries rarely thrive on heavy clay soils, fail on poorly drained ones and bear small crops on shallow or very light soils. Raspberries do best on deep, medium to light soils, particularly in higher rainfall areas or where the soil can be irrigated.

Isolating the Roots

One difficult problem with raspberries is that the roots will spread for many yards from the bushes and throw up suckers amongst the other fruit plots and outside the fruit cage. The only effective way of dealing with this problem is to place a physical barrier in the soil round the raspberry plot. This involves digging a narrow trench 2-3 ft (60-90 cm) deep, 3 ft (90 cm) from and across the ends of the raspberry rows. A sheet of 500 g polythene should be placed vertically against the walls of, but not in the bottom of, the trench. Polythene is the cheapest material but sheets of corrugated iron or a brick wall will serve the same purpose.

Planting Stock

It is more important with raspberries than with any other soft fruit to make sure that certified stock is being supplied by the nurseryman. The reason for this is that there has rarely been sufficient certified canes to supply both commercial growers and amateur gardeners. Consequently large numbers of poor quality virus infected canes are dug from old commercial plantations and sold to amateur gardeners. Only when gardeners stop buying these poor uncertified canes will more certified canes become available. It is essential to purchase canes that have been certified by the Ministry of Agriculture or Department of Agriculture for Scotland and to insist that the nurseryman should quote the certificate number of the stock that he is supplying.

Planting Distance

Many varieties of raspberry can be planted as close as 5 ft (1.5 m) between the rows, but varieties that have very long laterals, such as 'Admiral', 'Joy' and 'Leo', should be planted in rows that are 6 ft (1.8 m) apart. There is a choice of two 'in the row' planting distances. When the bushes are to be grown on the 'Scottish stool' system, the distance in the row should be 27 in (70 cm). For the English hedgerow system of production the distance should be reduced to 15 in (40 cm). (See Figure 8.1.)

The Scottish stool method is to be preferred as the canes grow as a discrete bush round which it is easy to kill weeds and surplus canes. With the English system, canes are allowed to grow in a continuous band 8-12 in (20-30 cm) wide along the row. It is very difficult to hoe weeds that grow amongst the canes, to restrict the number of new canes to the correct density and to prevent new canes from becoming infected by fungus diseases.

The planting canes should be pruned back to a height of 12 in (30 cm) before planting for either system and there is no advantage whatsoever in leaving this cutting back until after planting or until the spring.

Planting

The ideal time for planting raspberries is, in the UK, during November or December into soil that was dug sometime previously and allowed to settle. Planting can continue until March but growth may be more variable compared with that from earlier planting. The soil should be dry enough to allow it to be firmed onto the roots without puddling. Planting carried out later in the winter can be almost as satisfactory, provided the roots of the canes have not dried out and they go into moist soil.

A very satisfactory way of planting is to force a spade vertically into the soil to make a V-shaped slit. The roots should be pushed into the slit so that the uppermost roots are 2 in (5 cm) below ground and the tips of any white buds are at soil level. The slit should be closed on the roots by firming the soil back with the feet.

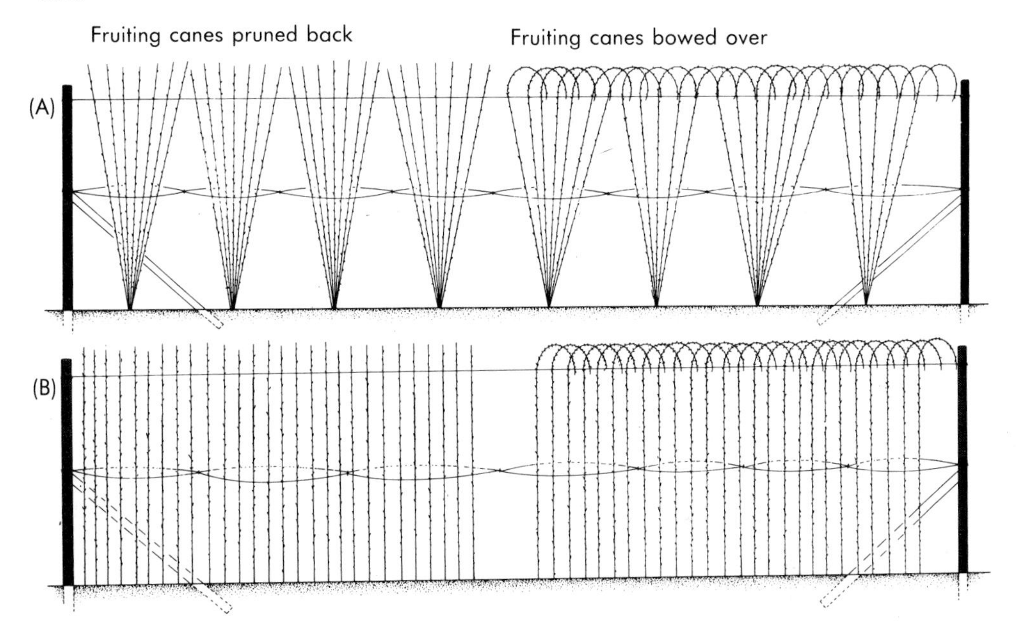

Figure 8.1: Raspberry Training Systems: (A) Scottish stool system; (B) English hedgerow

Manuring

At the end of March, broadcast on the soil in a circle 18 in (45 cm) in diameter round each cane:

2 oz (60 g) National Growmore fertiliser

or

2 oz (60 g) Phostrogen

or

1 oz (30 g) nitro-chalk
⅓ oz (10 g) sulphate of potash

Similarly, at the end of May, broadcast:

⅔ oz (20 g) nitro-chalk

In subsequent years, at the end of March, broadcast on the soil, 3 ft (1m) either side of the rows:

1 oz per sq yd (36 g/m^2) Phostrogen or National Growmore fertiliser

or

⅓ oz per sq yd (12 g/m^2) nitro-chalk
½ oz per sq yd (18 g/m^2) sulphate of potash

Posts and Wires

Posts and wires should be erected at the beginning of the winter after the first year's growing season and the canes tied to the wires. If tall canes are allowed to thrash around in the wind during the winter, they are likely to incur damage to their bases and die. Posts to support the wires are usually 6 ft 6 in (2 m) long, 2 x 3 in (5 x 7.5 cm) across and driven 1ft 9 in (52.5 cm) into the soil. Posts should be a maximum of 12 yds (11 m) apart and the end posts should be strutted. Galvanised 13 gauge wire should be used.

Traditional Training System

Two wires at a height of 24 in (60 cm) and another at a height of 54 in (1.4 m) are stapled loosely to the intermediate posts. The canes are secured between the bottom wires by tying the latter together at 3 yd (3 m) intervals and to the top wire by lacing with a continuous piece of twine or by individual ties (see Figure 8.2).

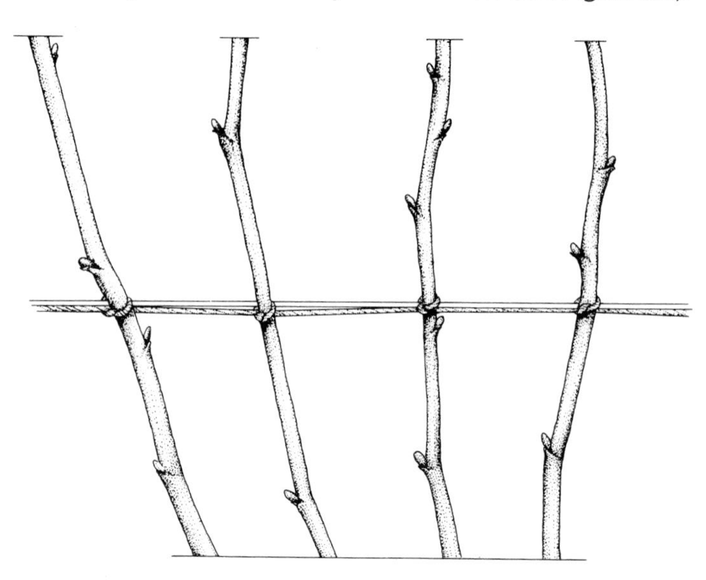

Figure 8.2: Raspberry Canes Laced to Wire with Polythene Twine

The Worcester System

This system was devised to separate the fruiting canes from the new canes and to make picking easier. Crossbars of wood 18-36 in long (45-90 cm) (depending upon how far the fruiting cane is to be separated from the new canes) should be bolted at their centres to the upright posts. The wires should run through staples that are driven into the ends of the crosspieces. Equal numbers of canes should be tied to each wire (see Figure 8.3). As many as 18 new canes per yd (20/m) should be allowed to grow, so that in their fruiting year they may be spaced as close as 3 in (7.5 cm) on the wires.

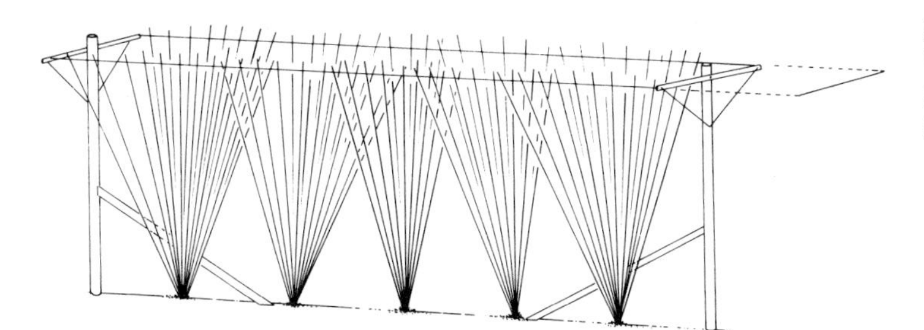

Figure 8.3: Worcester Training System for Growing Raspberries

Control of Numbers of New Canes

If maximum crops of fruit are to be grown each year, it is extremely important that the numbers of new canes should be rigidly controlled. If too many are allowed to grow, they will be drawn and spindly, the buds spaced too far apart, of poor quality and they may fail to break into growth in the following year. The canes will also become more easily infected with fungal diseases. Nine new canes per yard of row (10 per metre) is the maximum number that should be allowed to grow during the summer so that there should be a similar number to tie to the wires during the following winter.

If the Scottish stool system is adopted, it should be necessary only to destroy all the new canes that grow away from the original planted stool. This practice encourages the growth of strong good quality canes on the stools. Under the English hedgerow system it is more difficult to remove surplus canes and leave one cane in place for every 4 in (10 cm) of row (see Figure 8.1). Where the Worcester system of training is adopted a maximum of 18 canes per yard of row (20/m) should be retained (see Figure 8.3).

Whichever system is adopted the surplus canes should be either cut off with a sharp Dutch hoe or pulled out by hand before they reach a height of 6 in (15 cm). This should be done, in the UK, at the beginning of May, again in June and possibly in July. If by chance there are too many canes present in the row when the winter work is carried out, they should be dug out by the roots and not cut off at soil level with secateurs. If unwanted canes are cut off at soil level, a further two or more unwanted canes will appear in their place in the following summer.

Cane Vigour Control

Some varieties, and 'Glen Clova' is the best example, produce too many canes that grow too tall – over 7 ft (2 m) in height. This unwanted vigour can be reduced so that fruit quality is improved, yield increased and picking made easier. These objectives can be achieved by pulling or rubbing out every new cane that is present on the bushes when the average height of the new canes is between 4-8 in (10-20 cm) at the end of April or the beginning of May (UK). Within a fortnight, a replacement set of canes will grow away to take the place of those that have been removed, but they will be reduced in number and, as they are later, will not impede picking and their final height will be reduced.

Sucker Control

Raspberry suckers that grow up away from the rows do not seriously compete with the fruiting bushes but they are still a nuisance and have to be removed. They may be destroyed by rotavating between the rows but this will also destroy much of the effect of any weedkiller that was put on the soil during the winter. They may be cut off with any sharp tool such as a Dutch hoe or mechanically with a rotary grass mower. If the latter method is used, it is essential to remove any large stones and roll any small stones into the soil in order to prevent damage to the blades.

Weed Control

Raspberries are remarkably resistant to damage by simazine (Weedex) and it should be applied to both newly planted canes and established bushes. There is little risk of harming the canes if reasonable care is exercised in applying this chemical (see page 19). On newly planted canes, a slight ridge of soil 1-2 in (3-5 cm) high should be formed along the cane row before applying the Weedex. The rate of application should be:

Light soils – one sachet of Weedex in 2 gallons (9 l) of water to 75 sq yds (63 m^2)
Heavier soils – one sachet of Weedex in 2 gallons (9 l) of water to 50 sq yds (42 m^2)

On established bushes the rate of application should be:

Light soils – one sachet of Weedex in 2 gallons (9 l) of water to 50 sq yds (42 m^2)
Heavier soils – one sachet of Weedex in 2 gallons (9 l) of water to 33 sq yds (28 m^2)

To kill any seedling or established annual weeds, gramoxone (Weedol) should be added at the rate of one sachet to 1 gallon (4.5 l) of water and applied to 20 sq yds (17 m^2) of soil. Weedol is very safe for use amongst newly planted canes and established bushes during the winter whilst they are dormant. It may also be used during the summer as a carefully directed spray. Although spray drift will burn any leaves that are inadvertently sprayed, the stems of new and old canes will not be damaged.

Summer Work

Where the soil in which newly planted raspberries are growing has received an application of herbicide and no weed growth has occurred, the crop requires little attention during the summer.

Where a herbicide has not been applied, it is essential that the soil should be kept free from weeds by cultivating shallowly with a Dutch or draw hoe. Competition with weeds in early summer will seriously affect the growth of the raspberries.

The new cane growth, before and after it appears through the soil, should be regularly inspected as it can be attacked and killed by such pests as clay coloured weevil, leatherjackets or wireworm. The new shoots of vigorous varieties such as 'Glen Clova' and 'Delight' should appear through the soil before the end of April, whilst the shoots of shy varieties such as 'Malling Jewel' or 'Leo' may not emerge until a month later. Provided the new canes grow strongly, the flowers borne on the original cane may be allowed to bear fruit. Where shoot emergence is later than this the cause could be late planting, planting into a dry soil or mishandling of the canes before planting. If shoot emergence has been late and growth is poor, the short piece of fruiting cane should be cut back at soil level and not allowed to flower and bear fruit. When healthy, good quality canes have been cultivated and fertilised correctly, by the end of September the new canes should have reached an average height of 4 ft (1.2m).

Pruning

Summer fruiting raspberries may be pruned at any time after picking has stopped and as late as the following February. In one experiment, early removal in August of the spent fruiting canes did not improve the quality of the new cane or increase the weight of fruit they bore in the following year. However, it is an advantage to prune in August or September when the new canes are to be woven on to the wires as the canes are more supple at this time. This method makes tying with string unnecessary.

The procedure should be to cut out with secateurs each spent fruiting cane, leaving the smallest stub possible. Then dig out or pull out by their roots all new canes that are growing away from the stools and, in the case of the hedgerow system, every cane that is situated at a distance of more than 6 in (15 cm) from the row. Next, cut off any spindly, short or damaged canes that are growing on the stools and in the hedgerow system similarly dig or pull out individual spindly, short or damaged canes. If, after this, there is still an excessive number of sound tall canes remaining in the row, the number should be reduced to a maximum of eight per stool, or nine per yard (10/m) of row, or 18 per yard (20/m) for the Worcester system of training.

Once pruning has been completed, it is advisable that the canes should be secured to the wires to prevent them thrashing about in the wind and being damaged. They may either be tied between pairs of bottom wires, tied individually, laced to or twisted on to the upper wire(s) (see Figure 8.2).

Raspberry canes frequently grow 7 ft, 8 ft and 9 ft (2.1 m, 2.4 m and 2.7 m) tall and it is on the upper 2 or 3 ft (60 or 90 cm) that the better quality buds are borne. To avoid cutting off these buds, the canes should either be bowed over in a semi-circle and the cane tied a second time at its tip to the wire (see Figure 8.1), or if this does not accommodate all the cane, a further measure should be to train the canes to the wires at an angle of 60°. The weaving of long canes along the upper wire makes pruning unnecessary but this method gives rise to a crowded mass of fruiting laterals that become drawn and easily broken. Whichever method of training is adopted, the final operation is to prune off the weak 6 in (15 cm) tip of each cane on which buds of poor quality are borne. Where canes grow to an excessive height in the following years, the application of fertiliser should be reduced or omitted and cane vigour control practised in May (see p. 100).

Cultivation of Autumn Fruiting Varieties

An autumn fruiting variety is one that flowers on the current season's new cane and the fruit of which ripens from August or September onwards (UK). If the canes are allowed to grow in the following year, the dormant buds on the lower half of the canes will flower and bear fruit in mid-summer. In order to maximise the amount of autumn fruit, all the canes are cut down to the ground during the winter and the summer crop is not taken. As the comparatively small amount of fruit that is ripening at any one time is very attractive to birds, it is probably advisable for autumn fruiting raspberries to be grown in fruit cages.

Currently available varieties are probably only worth growing in the southern half of the United Kingdom, in sheltered gardens, if any quantity of sound ripe fruit is to be picked. In the near future, the introduction of 'Autumn Bliss' – a new earlier ripening variety – should enable successful production to be extended further north.

Autumn fruiting raspberries should be cultivated in a similar manner to summer ripening varieties. However, there are several important differences. Single canes should always be planted 15 in (45 cm) apart in the row and the distance between the rows should never be less than 6 ft (1.8 m) apart. The new canes that grow in the first and subsequent years should be allowed and encouraged to appear in a band 2-3 ft (60-90 cm) wide. Any canes that appear outside the band should be rigorously suppressed whilst still small by cutting or hoeing off two or three times during the summer. Every winter the canes that have borne fruit should be cut back to soil level.

It is almost impossible to weed amongst the band of canes without damaging them; it is therefore essential that the soil should have been freed completely from perennial weeds before planting. Annual weeds must be controlled, each February (UK), by applications of Weedex.

Support

It is advisable to provide the canes with support so that they do not bend over excessively with the weight of fruit. A support also helps to prevent wind damage due to the berries rubbing against leaves and canes.

The best method is to drive 4 ft (120 cm) long, 2½ in x 2½ in (6 cm x 6 cm) stakes into the corners, as well as the sides of, the beds at 11 yd (10 m) intervals. When the canes have attained a height of 3 ft (90 cm), strings should be secured to the posts (along the sides of the bands of canes). Better protection will be afforded if a polypropylene net with a 6 in (15 cm) mesh is stretched between and secured to the posts with string in June, just before the canes attain a height of 3 ft (90 cm).

Watering

Soil in which raspberries are growing should be brought up to its full water-holding capacity with an application of 9 gallons per sq yd (50 l/m^2) shortly before the berries turn pink. Excessive applications of water to raspberries makes the canes grow too tall and liable to infection by fungal diseases. Raspberries that have not grown to a satisfactory height in the previous year or are growing on shallow light soils may be watered before the end of June.

Picking

Raspberries are in a prime condition for picking and immediate consumption when each variety has developed its characteristic colour. 'Admiral' and 'Malling Jewel' turn a deep red when fully ripe whilst 'Glen Clova' and 'Leo' are deep pink and 'Delight' is pale pink in colour. When temperatures are in the seventies (20°C) the bushes should be picked over every day, preferably in the morning whilst the berries are still cool. Under cool weather conditions (50-60°F; 10-15°C) the rate of ripening slows down and picking may only be necessary every second or third day. Ripe berries will keep for 48 hours in a refrigerator. If they are picked after just having turned pink they may be stored in this way for seven days. Fully ripe fruit is best for deep freezing and should be placed in a single layer on a tray in the freezer; after four or five hours when they have frozen hard they should be transferred to platic bags for permanent deep frozen storage.

Pests and Diseases

Raspberries should always be sprayed as a matter of routine to control raspberry beetle and in the wetter parts of the United Kingdom for botrytis grey mould. It is not absolutely advisable or necessary in the eastern drier parts of the country to spray for the latter. Sprays for the other pests and diseases should only be applied if they are expected or are present on the bushes.

Propagation

Provided that a certified stock of raspberries was planted, in particular a variety that is resistant to aphid colonisation, it would not be unreasonable to propagate from such a stock in the second or third winter after planting. The raspberry is the easiest of crops to propagate. Any new cane that is not attached to the stool when dug up with a spade will have sufficient root for it to transplant successfully and grow vigorously during the following summer.

The raspberry may also be propagated by root cuttings. In early winter use a spade to dig up the roots of the bush selected for propagation. Thin and thick roots should be cut into 4-8 in (10-20 cm) lengths. Two pieces of root should be planted together in prepared soil at a depth of 2-3 in (5-7.5 cm) in the positions where it is wished to establish bushes. Alternatively, two pieces of root may be placed 2 in (5 cm) deep in a 4 in (10 cm) pot of compost. Young shoots will emerge from the pots and they can be planted out anytime during the following summer, provided they can be watered until the roots establish themselves in the soil. If it is not convenient to plant out the young shoots into the open ground during the summer, they can be potted on into somewhat larger pots, and kept growing in this way until the following winter when they should be planted into the soil in their permanent position. The young canes should not be cut back but allowed to bear fruit.

Table 8.2: Raspberry Protection Chart

Problem	Chemical	Key	Product
Aphids	Dimethoate	1	Boots Greenfly & Blackfly Killer
		6	Murphy Systemic Insecticide
	Fenitrothion	7	PBI Fenitrothion
	Malathion	7	Malathion Greenfly Killer
		6	Murphy Liquid Malathion
	Permethrin	7	Bio Sprayday
		2	Fisons Whitefly & Caterpillar Killer
		3	Picket
	Pirimiphos-methyl	3	Sybol 2
Botrytis Grey Mould	Benomyl	3	Benlate plus Activex
	Carbendazim	1	Boots Garden Fungicide
	Thiophanate-methyl	4	Fungus Fighter
		6	Murphy Systemic Fungicide
Cane Spot	Benomyl	3	Benlate plus Activex
Cane Botrytis	Carbendazim	1	Boots Garden Fungicide
Spur Blight	Thiophanate-methyl	4	Fungus Fighter
		6	Murphy Systemic Fungicide
Green Capsid Bug	Malathion	7	Malathion Greenfly Killer
		6	Murphy Liquid Malathion
	Fenitrothion	7	PBI Fenitrothion
	Permethrin	7	Bio Sprayday
		2	Fisons Whitefly & Caterpillar Killer
		3	Picket

March April May June July Aug. Sept.

Pest/Disease	Active ingredient		Product
	Pirimiphos-methyl	3	Sybol 2
Clay Coloured Weevil	Fenitrothion	7	PBI Fenitrothion
	Pirimiphos-methyl	3	Sybol
	Carbaryl	1	Boots Garden Insect Powder
Leatherjackets	Diazinon	6	Rootgard
	Phoxim	2	Fisons Soil Pest Killer
	Pirimiphos-methyl	3	Sybol 2
Powdery Mildew	Benomyl	3	Benlate plus Activex
	Carbendazim	1	Boots Garden Fungicide
	Bupirimate & Triforime	3	Nimrod T
	Sulphur	5	Comac Cutonic Sulphur Flowable
Raspberry Beetle	Fenitrothion	7	PBI Fenitrothion
	Malathion	1	Malathion Greenfly Killer
		6	Murphy Liquid Malathion
	Rotenone	7	Liquid Derris
Raspberry Cane Midge	Fenitrothion	7	PBI Fenitrothion
Raspberry Moth	Tar Oil	3	ICI Clean Up
		6	Murphy Mortegg
Red Spider Mite	Malathion	7	Malathion Greenfly Killer
		6	Murphy Liquid Malathion
	Pirimiphos-methyl	3	Sybol 2
	Dimethoate	1	Boots Greenfly & Blackfly Killer
		6	Murphy Systemic Insecticide
Tortrix Moth Caterpillars	Fenitrothion	7	PBI Fenitrothion
	Malathion	7	Malathion Greenfly Killer
		6	Murphy Liquid Malathion
	Permethrin	3	Picket

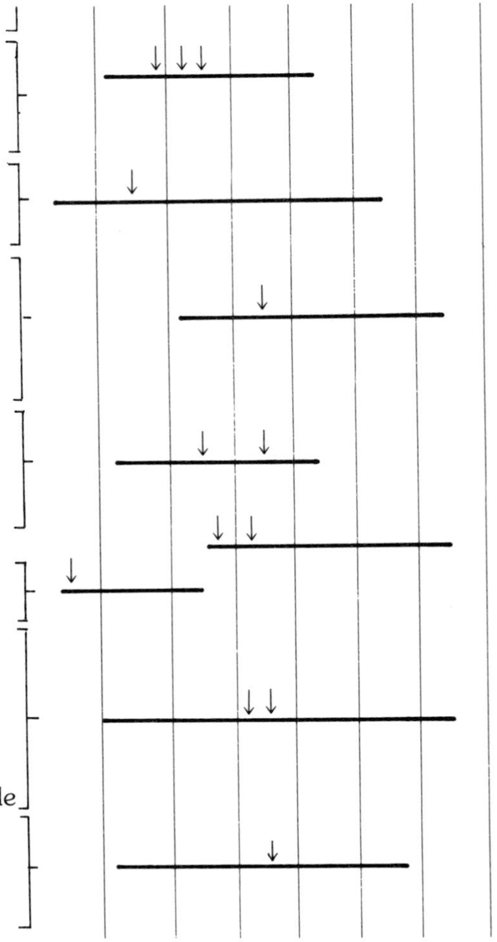

Problem	Chemical	Key	Product	March	April	May	June	July	Aug.	Sept.
Wireworm	Pirimiphos-methyl	3	Sybol 2	—	↓ —	—	—	—	—	
	Phoxim	2	Fisons Soil Pest Killer							
	Diazinon	6	Root Guard							

Key: ——— =Infestation periods ↓ = Timing of pesticide application

1 The Boots Company PLC, Nottingham.
2 Fisons PLC, Paper Mill Lane, Ipswich.
3 ICI Products, Woolmead House, Farnham, Surrey.
4 May & Baker, Regent House, Hubert Road, Brentwood, Essex.
5 McKecknie Chemicals Ltd, PO Box 4, Widness, Cheshire.
6 Murphy Chemical Co., Latchmore Court, Brand Street, Hitchin, Herts.
7 Pan Britannica Industries, Waltham Cross, Herts.

Chapter 9 Strawberries

Introduction

The attractions of strawberries are many; the cost of planting stock is small, and they produce fruit even when given the minimum of attention and expense, though under these conditions the crop will vary considerably from year to year. The berries have a more attractive colour than any other fruit, are sweet and well flavoured, enabling them to be eaten when picked direct from the bush – though they are even better when eaten with sugar and cream. Their one fault is that they do not preserve very well and this provides every incentive to prolong the ripening season for as long as possible, by various means. They grow best in zones of hardiness 5-8.

The wild strawberries (*Fragaria vesca* and *F. elatior*) grow in woods and hedgerows in Europe. The introduction of the Virginian strawberry (*Fragaria virginiana*) from North America and the Chilean strawberry (*Fragaria chiloensis*) from South America, with their better flavour and size respectively, enabled breeders to develop today's heavy cropping, well flavoured, large fruited varieties. A variant of our native strawberry (*Fragaria vesca semperflorens*) gave rise to the so-called 'perpetual' fruiting type. Varieties bred from this source produce ripe fruit throughout the summer and autumn but the berries are smaller than those of the summer fruiting varieties. Breeders are now concentrating on producing larger fruited and heavier yielding 'perpetual' fruiting varieties.

Varieties

There are more than 60 varieties of strawberry under cultivation in Europe. These cannot all be worth growing in the United Kingdom and a selection has been made of those varieties that are well established and of new varieties that have been raised in continental countries, North America and in the UK.

Anyone who is growing strawberries for the first time should select 'Cambridge Favourite' because it is the most reliable heavy cropping variety and succeeds in gardens all over the country. To obtain the earliest ripening fruit outdoors, select 'Redgauntlet' and to extend the season, 'Domanil'. For those who wish to experiment, 'Hapil' and 'Tamella' are better flavoured maincrop varieties; 'Tenira' is a late-ripening variety of excellent flavour. The early-ripening varieties 'Cambridge Vigour', 'Grandee' and 'Pantagruella' are all better flavoured than 'Redgauntlet'. 'Maxim', a new variety from Holland, became available for the first time in 1983 and produces fruits that are even larger than those of 'Grandee'. The British varieties 'Saladin', 'Tantallon' and 'Troubadour' are resistant to many of the soil diseases that trouble strawberries, particularly where the crop has to be grown without an adequate rotation.

For the production of late fruits from August onwards, there are now seven 'perpetual' fruiting varieties from which to choose. 'Aromel' is the better flavoured and 'Ostara' the heavier cropping variety. 'Marastar' is a new variety said to be resistant to mildew that has yet to prove itself in this country.

Flavour

Flavour is so much a matter of personal taste; the flavour of each variety is different in subtle ways and it is difficult, if not impossible, to convey in a meaningful way the differences between them. The flavour of each variety has been classified in this book as poor, fair, moderately good, good or very good, representing generally held opinions about these varieties.

Early Mid-summer Varieties

'Cambridge Vigour' (UK)

This variety can crop as heavily as 'Cambridge Favourite' but size of fruit and yields deteriorate on light soils under drought conditions. The fruits are large in the first cropping year but usually smaller in later years. For this reason it is usually only cropped for two years. The fruits ripen early, are conical in shape, bright shiny scarlet in colour and fairly firm-fleshed. It is susceptible to verticillium wilt and even more so to mildew and should always be treated regularly for these diseases. Early autumn planted maidens produce several trusses of fruit in the following season. Flavour very good.

'Gorella' (The Netherlands)

'Gorella' is a second early variety that bears good crops of conical berries, though cropping from one year to another can be somewhat erratic. The berries are crimson-red in colour and the tips, particularly those of the larger berries, remain green unless they are allowed to ripen fully before picking. The flesh is bright red and of moderate flavour. The habit of the plant is compact with tough leathery leaves, which are susceptible to powdery mildew.

'Grandee' (West Germany)

This early-ripening variety is outstanding because of the large size of its fruit that can measure as much as 3 in (7.5 cm) in diameter and 3 oz (85 g) in weight. It also crops heavily, especially in its second year. It is somewhat susceptible to post-harvest mildew and fungicide should be applied to control this disease. The variety is much used to win fruit size competitions. Flavour fair.

'Idil' (Belgium)

This new variety should be on sale for the first time in 1985. It is somewhat similar to 'Cambridge Favourite', bearing heavy crops, and the berry separates very easily from the calyx. The fruits are medium-sized, round-conical in shape and bright crimson in colour. Flavour good.

'Pantagruella' (West Germany)

The earliest of all varieties, and producing a good crop of fruit for such an early variety. Extremely compact plants that should be planted as close as 9 in (22.5 cm) apart in the row and grown as single plants. On account of its earliness, plants may require frost protection during the flowering period. An ideal variety for covering with cloches or polythene, giving a maximum crop when shop prices are extremely high. The attractive medium-sized conical fruits with orange-red flesh tend to be small unless watered. It is liable to infection by mildew after harvest. Flavour good.

'Redgauntlet' (UK)

This is a heavy cropping variety that bears large to medium, round, bright red fruits that may be 'cat-nosed' when pollinating conditions are poor. The maiden crop is early but the fruit from older beds ripens somewhat later. 'Redgauntlet' does well under cloches and tunnels and, in the southern parts of the United Kingdom, invariably bears a second crop in late summer. In hot summers, plants grown in the open also produce a second crop. It is resistant to mildew and grey mould. Flavour fair.

Midsummer Varieties

'Cambridge Favourite' (UK)

This is the most widely grown variety by commercial and private gardeners because it can be relied upon to crop well under a wide range of conditions; also the fruits can be left to 'hang' on the plants for a long period without getting over-ripe or going rotten. The berries are round to conical in shape and pale red in colour with firm white flesh. They are medium in size and this is maintained throughout the season. It is very susceptible to infestations by red spider mite but is resistant to infection by mildew. Flavour moderately good.

'Hapil' (Belgium)

This is a comparatively recent introduction that gives good yields of large, conical bright red fruits. It may not be as adaptable to local conditions as 'Cambridge Favourite' but it is a variety that should be given a trial, for where it does well, it is a better variety than 'Cambridge Favourite' to grow. It crops well on light soils and under dry conditions, but is susceptible to verticillium wilt and red spider mite. Flavour very good.

'Harvester' (UK)

Harvester is a new variety with yields that vary between moderate and heavy. The fruits are orange with orange flesh, firm but small; a high proportion of the fruits may be malformed. Flavour poor.

'Korona' (The Netherlands)

A new heavy cropping variety which ripens about the same time as 'Cambridge Favourite', but having much larger fruits than that variety. They are firm and juicy, fairly dark red and of even size. The flesh is red. It is slightly susceptible to mildew. Flavour good.

'Maxim' (The Netherlands)

A new variety first introduced in 1983. 'Maxim' will produce large single wedge-shaped medium red fruits that are as large or larger than those of one of its parents, 'Grandee'. Very heavy cropping and resistant to drought. Flavour good. (See Plate 13.)

'Royal Sovereign' (UK)

This variety is the only one that has survived the Second World War and is said by some to be of outstanding flavour – a view which is now being challenged. The berries vary in size from small to large, are wedge-shaped and orange-red in colour. It is low yielding and beds have a short life because it becomes rapidly infected with virus diseases and is susceptible to all the other troubles that affect strawberries. Flavour very good.

'Saladin' (UK)

This new late heavy cropping variety bears medium-sized fruits that are long, conical or wedge-shaped but liable to be badly misshapen. It is highly resistant to red core disease and has some resistance to grey mould and mildew. The berries are bright red with pale red flesh. The variety is prone to stress in drought conditions. Flavour moderate.

'Silver Jubilee' (UK)

The open habit allows the fruit to be easily seen but the yield is less than 'Cambridge Favourite'. The berries are long, conical in shape, light red to pink in colour, with pale rose-coloured flesh but liable to be misshapen. Flavour moderately good.

'Tamella' (The Netherlands)

Maiden year season is early. After the first year of fruiting it reverts to mid-season/late, cropping over a very long period. 'Tamella' gives quite a useful crop in its maiden year even with late autumn planting and succeeding crops in following years are extremely heavy. The attractive looking berries maintain their yield and size even during the third year of fruiting and 'Tamella' will outyield all other commercial varieties. The berries are very large, conical or wedge-shaped and they have to be carefully handled as they bruise easily. This variety is susceptible to crown rot on some soils. Flavour good.

'Tantallon' (UK)

This variety has the potential to outyield 'Cambridge Favourite', provided it is supplied with enough moisture. It is a moderately vigorous variety, so that much of the fruit is partially hidden by the leaves. The fruit is glossy orange-red, wedge- to conical-shaped with light red flesh that darkens with over-maturity. The plants are prone to stress in dry summers. It is field resistant to red core and has some resistance to grey mould and mildew. Flavour moderately good but rather acid.

'Totem' (Canada)

The outstanding characteristic of 'Totem' is that when defrosted after freezing the fruit has a better shape retention, colour and texture than other varieties. The fruit should be frozen sliced or whole, with sugar syrup. 'Totem' is a moderate cropping variety that bears medium-sized fruits which turn dark red when fully ripe and has dark red flesh. The variety is not especially recommended for fresh fruit consumption. Flavour good.

Late Mid-summer Varieties

'Bogota' (The Netherlands)

This is one of the newer varieties and still undergoing trial. It is worth growing because it is very late and may extend the strawberry season still further (beyond 'Domanil'). The berries are medium to large in size, conical or wedge-shaped with a dull red colour. It should bear moderate, if not heavy, crops. It is somewhat susceptible to infection by mildew. Flavour moderately good.

'Domanil' (Belgium)

This is the latest ripening mid-season variety generally available at present and is much in demand by professional fruitgrowers. The berries are large, wedge-shaped, sometimes ribbed, with dull, moderately-dark colour. A heavy cropper. Flavour moderate.

'Providence' (LARS)

This is a new variety that is unlikely to be available for planting until 1988. The berries are a pale red colour similar to 'Cambridge Favourite', medium-sized and a high yielding variety with a good flavour.

'Tenira' (The Netherlands)

'Tenira' has bright crimson, medium-sized conical-shaped fruits. The flesh is mid-red and has a moderately firm texture. 'Tenira' is capable of bearing heavy crops of fruit when moisture and soil conditions are ideal and has the reputation of being the best flavoured modern variety. In its maiden year it can outyield 'Cambridge Favourite' but in succeeding years crops may be much less. Like 'Cambridge Vigour', it is best grown for two years only and then replanted. In some years, the variety bears a second crop. Flavour very good.

'Troubadour' (UK)

A very vigorous late mid-season variety, the leaves of which hide most of the fruit. The medium-sized conical or wedge-shaped fruits are vermillion-red in colour with medium red flesh. It is fairly resistant to many diseases, including verticillium wilt. The variety does not thrive under drought conditions. Flavour fair.

12 Raspberry: 'Glen Moy' Variety (photo: SCRI)

13 Strawberry: 'Maxim' Variety

14 American Gooseberry Mildew on Blackcurrant

15 Botrytis Grey Mould Infection on Strawberry

16 Lesions of Canker (*Godronia cassandrae*) on Blueberry (photo: SCRI)

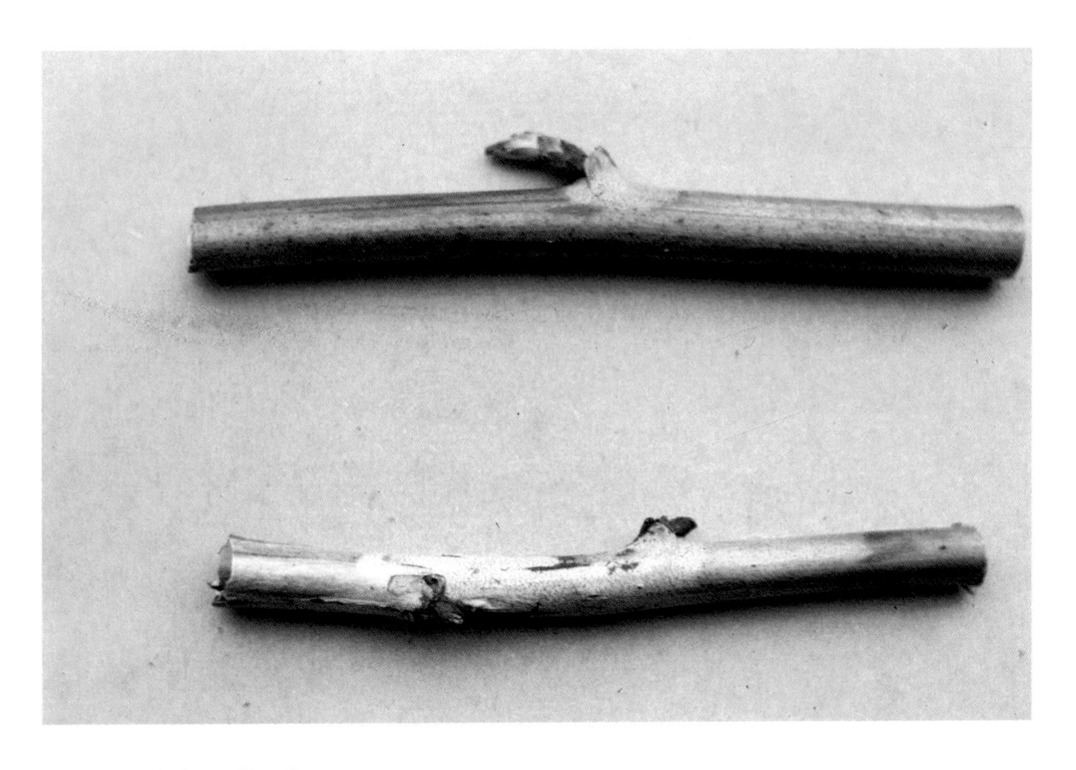

17 Spur Blight on Raspberry

18 Currant Sowthistle Aphid on Blackcurrant

19 Big Bud Mite on Blackcurrant

20 Sawfly Caterpillar on Gooseberry

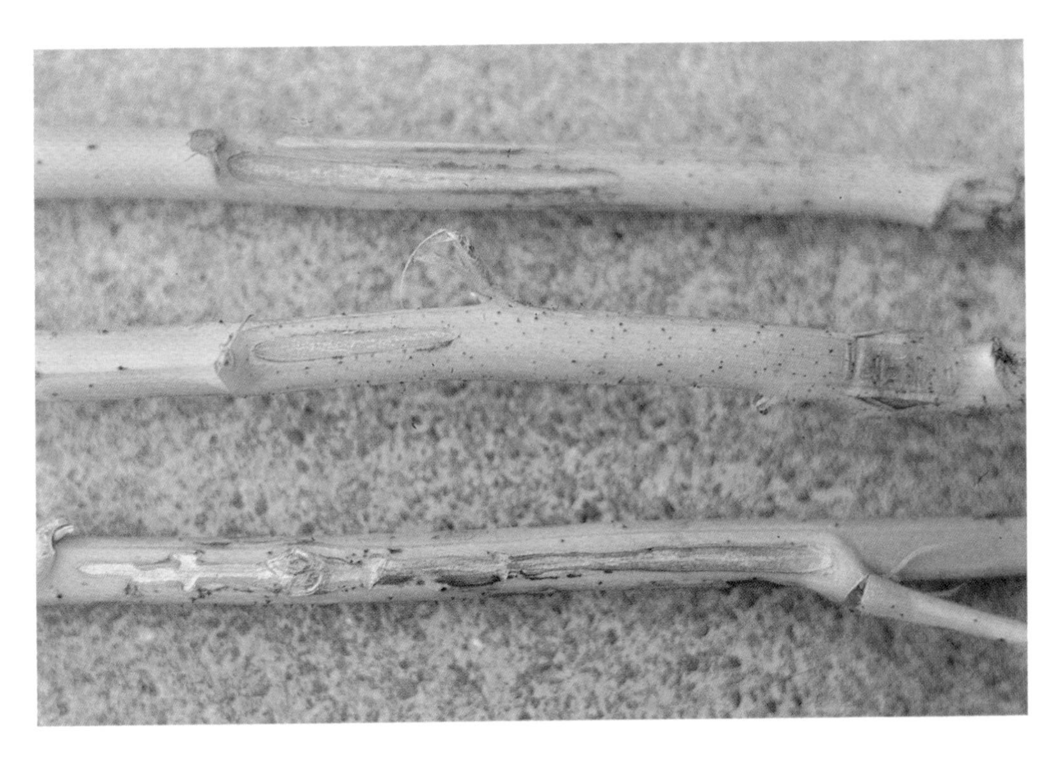

21 Raspberry Cane Midge: Egg Laying Sites on Raspberry

22 Red Spider Mite on Blackcurrant

23 Flowers Killed by Strawberry Blossom Weevil

'Perpetual' Fruiting Varieties

'Aromel' (UK)

An autumn fruiting variety bred in Dorset by Richard Cumberland. As a 'perpetual' fruiting variety, 'Aromel' crops at similar levels to 'Ostara' and 'Rabunda'. The fruit is medium to large, conical or short wedge-shaped, with medium red flesh. Runner production is usually better than most 'perpetual' varieties. A useful fresh fruit variety for the August to October period, best grown for one maiden crop and then replanted. Flavour very good.

'Fraise du bois'

A non-runnering selection of the Alpine strawberry. It produces an abundance of thimble-sized fruits throughout the summer and early autumn. It makes a splendid border plant but grows best in a shaded situation. Flavour good but aromatic.

'Gento' (West Germany)

This variety produces larger fruits than the other 'perpetual' fruiting kinds, probably because it bears a summer and autumn flush of fruit, rather than over the whole season. 'Gento' crops freely on its runners and will grow well on chalky soils. The colour of the fruit is reddish-crimson with red flesh. Even when planted in the early spring, a full crop can be expected in the same year, provided the plants have sufficient moisture throughout the growing season. It is susceptible to mildew. Flavour good.

'Marastar' (France)

This new variety is the first 'perpetual' fruiting strawberry which is said to be resistant to mildew, so that the control of the disease by spraying appears to be unnecessary. 'Marastar' also has some resistance to botrytis. The plant has a spreading habit, so that the flowers and fruit are well displayed. The berries are medium-sized with a medium-red skin and partly coloured flesh. In France, 'Marastar' has cropped heavily for its type. Flavour good.

'Ostara' (The Netherlands)

A good cropping 'perpetual' fruiting variety that crops very satisfactorily in late summer. The fruits are medium in size, of round or conical shape and well displayed, with red skins and orange flesh. The variety is susceptible to mildew. Flavour good.

'Rabunda' (The Netherlands)

Crops reasonably well for a 'perpetual' fruiting variety. The fruits are medium-sized round-conical, orange-red in colour with pale orange flesh. The variety is susceptible to mildew. Flavour moderate to poor.

Planting Stock

Although it is extremely important that only stocks of strawberry varieties that have been certified for apparent freedom from virus and other diseases should be planted, this is not always possible; the United Kingdom Nuclear Stock Associations produce only virus tested plants of varieties that are in demand by *commercial growers*. This means that many of the older varieties and some varieties of continental origin cannot be purchased with a Special Stock or 'A' certificate. The position will be improved somewhat by the granting of an 'H' certificate to these varieties. Plants with an 'A' certificate are perfectly satisfactory for planting in private gardens. Failing this, plants with an 'H' certificate should be purchased.

Time of Planting

Strawberries may be planted at any time when the soil is in good condition between February and November in the southern parts of the United Kingdom and between March and October in the north of the country. Planting should not be carried out during the winter periods because the plants will not root into cold soil and hard frosts will lift them out of the ground. If runners are purchased during this period they should be potted up and kept in a cold frame until they can be planted in the spring.

Effect of Planting Dates on Yield

The time of planting has an important bearing on the amount of fruit that will be borne by the plants in the following maiden year. The earlier the planting the higher the yield will be. Early planting is also an insurance for a high yield in the first full cropping year and allows time for any inadvertant check to be overcome.

Maiden Year.

Time of Planting	Type of Plant	Yield per Plant
Early July	Cold-stored runners	16 oz (450 g)
August-September	Runners in pots	8 oz (227 g)

October-November	Fresh dug runners	4oz (113 g)
Spring planting	Fresh dug runners	2 oz (57 g)

First Cropping Year.

Single-spaced plants (15 in (37.5 cm) apart)	16 oz (450 g)
Matted rows	3 lb per yd (1.4 kg/m)

(A Jiffy 7s is compressed peat compost contained within a nylon mesh. When soaked in water, it swells up to 7 times its original thickness, when it measures 2 in (5 cm) in diameter and 1½ in (3.75 cm) in depth.)

Types of Plant

Freshly Dug Runners

Freshly dug runners of good quality should be available for planting in October and November and again in the spring from February to April. The quality of a runner should be judged solely by its root system. This should consist of ten primary roots each 4 in (10 cm) in length and furnished with laterals. A runner as described would have four to six leaves and roots that have not been allowed to dry out. Plants in pots should not have a starved appearance and the roots should have spread throughout the compost.

Cold-stored Runners

Cold-stored runners should be of identical quality to open-ground runners but they will have had their leaves cut off by the nurseryman. They would have been lifted during the previous months of December or January and stored at 30°F (−1°C) until despatched. They are available for purchase from May until mid-August and should be planted and watered-in within 24 hours of receipt. If they are still frozen or are still cold when received they could be kept in a refrigerator at 32°F (0°C) for up to 48 hours. Planting should be completed by early July in the north, by late July in the south and early August in the south-west parts of the United Kingdom. Planting after these times may lead to some of the plants failing to flower in the following year.

Runners growing in Jiffy 7s, Small Pots and Peat Blocks

Established runners growing in Jiffy 7s, small pots or peat blocks, are usually available for planting in the period August, September and early October, a period during which open-ground plants have small crowns and may be poorly rooted.

Such plants are produced by rooting cold-stored runners or stolon plantlets in Jiffy 7s, pots or peat blocks and growing them in a glasshouse until they are well rooted. Cold-stored runners and plants in pots rarely fail to establish well, although they are expensive compared with freshly dug runners.

Soil Preparation

With strawberries it is particularly important that the soil should be free from perennial weeds before planting and that the procedures recommended on page 6 should have been carried out. It is impossible to get rid of perennial weeds in an established bed and trying to do so causes severe root damage to the fruit plants. It is very important to fork up the sub-soil to make sure the soil is not panned in any way and to provide free-draining soil conditions that do not favour infection of the roots with red core disease. The soil should have been dug as soon as possible after the previous crop was removed from the ground – if possible several weeks before the strawberries are to be planted, so that the soil has time to settle and provide a good planting bed.

If possible, after digging, farmyard manure or compost should be spread over the soil at the rate of one to two barrowloads per 10 sq yds (8.5/m^2). It should be mixed well with the soil by forking or rotavating. Immediately before planting, broadcast over the soil one of the pre-planting fertilisers listed on page 12. The fertiliser should be raked into the soil. If necessary, the soil should be raked and rolled or trampled in succession until it is as firm as a seedbed required for the sowing of vegetable seeds.

Planting Distances

The distances at which, in the past, strawberries have been planted were as close as 27 or 30 in (70-75 cm) between the rows and 12-15 in (30-37 cm) between the plants in the rows. These distances are really too close and should only be used in small gardens where space is at a premium. At these distances it is difficult to avoid treading on the fruit and the plants compete too much with each other in the row. The one advantage of close planting is that in the maiden year it gives a relatively high yield per square yard. 'Pantagruella' and 'perpetual' varieties should be planted 9-12 in (22-30 cm) apart in the row because they do not grow as vigorously as other varieties.

Where space is not a limiting factor the healthy, more vigorous varieties should be planted 33-36 in (82.5-90 cm) between the rows and 12-18 in (30-45 cm) between the plants. For single-spaced plants, planting should be 12-15 in (30-37 cm) and for matted rows (see p. 124), planting should be 15-18 in (37-45 cm).

Treatment of Plants

On receipt, parcels of plants should be examined immediately and returned to the nurseryman if they are not of a satisfactory quality. If found to be satisfactory they should be taken out of their bundles and, whether bare-rooted or in pots, placed in water for a few minutes. If they cannot be planted immediately, they should be heeled in a single line in moist soil or peat in a shady place.

Planting

Planting should be carried out with a spade or trowel with which a V-shaped opening should be made in the soil. The crown of the strawberry runner should be placed on the edge of the slit with the roots hanging on its face. The slit should then be closed by pushing the soil back with the planting tool and well firmed on the roots with one's boot (see Figure 2.4). When the operation has been completed, the runner should be so firmly anchored in the soil that if an attempt is made to pull it out of the soil when holding one of the leaves, the leaf should tear before the plant moves. It is also very important that the crown of the runner should be just sitting on the soil surface so that it is neither buried nor with any roots exposed. Runners that are planted too deeply or with roots exposed above the soil rarely thrive and frequently die.

Weed Control

A strawberry bed may be kept free from weeds throughout its life by hand hoeing but this is tedious work and difficult to do properly during wet weather. Hoeing can also damage the roots of the strawberry plants and, when grown in matted rows, interferes with rooting of runners. This work should be carried out using a Dutch or swan-necked hoe, taking care not to drag soil away from the plants or allow any weeds to grow past the seedling stage of development.

There are now three chemicals available in the United Kingdom that, if used with care, enable strawberry beds to be kept free from weeds for long periods. Because the soil and roots of the strawberry plants remain undisturbed, the yield of fruit will be heavy, if not heavier than, a comparable bed that has been kept free from weeds by hand cultivation.

Covershield

Immediately after planting, when the soil should be weed-free and have a fine firm tilth, Murphy Covershield should be broadcast over the soil with the shaker that is supplied with the herbicide:

1 oz (28 g) Covershield over 8 sq yds (6 m^2)

The Covershield prevents weed seed germination for six to eight weeks, when further applications should be made.

In the month of August, following autumn or spring planting, Covershield should be replaced by Weedex (simazine), as the latter has a longer lasting effect on the weeds – nearly six months.

Weedex

Weedex should only be applied if the strawberries are growing strongly and have produced at least four fully expanded healthy leaves. Before applying the Weedex, the bed should be absolutely free from weeds and the soil should have a moist, firm tilth. There is a choice of applications.

One application that should be applied each year in August after harvest:

Light soils – one sachet of Weedex in 2 gallons (9 l) of water to 50 sq yds (42 m^2)
Heavy soils – one sachet of Weedex in 2 gallons (9 l) of water to 33 sq yds (27 m^2)

or

Two applications each year, one made in August after harvest and the second in November:

Light soils – one sachet of Weedex in 2 gallons (9 l) of water to 100 sq yds (84 m^2)
Heavy soils – one sachet of Weedex in 2 gallons (9 l) of water to 66 sq yds (55 m^2)

The two-application method is the better one as it reduces the risk of the herbicide damaging the plants and maintains the concentration of herbicide in the soil at a higher level for a longer period of time.

The herbicide should be applied very carefully in accordance with the instructions on page 19, so that the correct dose is applied evenly over the bed. Weedex should *not* be applied to very light sandy or gravelly soils.

Weedol

Weedol is a contact herbicide that kills all green plant matter with which it comes into contact, but is inactivated and becomes harmless when it falls on the soil. It should be used in strawberry beds to kill weeds growing between the rows, and surplus runners that root between the rows. Weedol is best applied with a watering can and dribble bar. It can also be applied with a spraying machine operating at low pressure and with a large aperture nozzle to provide a coarse spray that is less likely to drift onto the strawberry plants.

One sachet of Weedol in 2 gallons (9 l) of water should treat 20 sq yds (17 m^2)

To kill strawberry runners, two applications of Weedol should be made, the first at the end of August or beginning of September when half the rows between the plants will be covered with runners and again at the end of November when the runners have stopped growing (UK). Before making the applications of Weedol the stolons that connect the runners to the parent plants should be cut. This should be done with a sharp spade or turfing iron, cutting down single lines on both sides of and 8 in (20 cm) from the parent plants. Great care should be taken to ensure that no Weedol falls outside the two lines of cuts.

Manuring

The fertiliser that should be broadcast over the surface of the soil before planting strawberries should be:

½ oz per sq yd (18 g/m^2) nitro-chalk
2 oz per sq yd (70 g/m^2) superphosphate
⅔ oz per sq yd (24 g/m^2) sulphate of potash

or

2 oz per sq yd (70 g/m^2) Growmore

or

2 oz per sq yd (70 g/m^2) Phostrogen

After the establishment year has passed, in every succeeding February, broadcast over the bed:

⅔ oz per sq yd (24 g/m^2) sulphate of potash

No other application of fertiliser should be made unless the vigour of the bed seriously declines. As soon as this is seen to occur, apply:

½ oz per sq yd (18 g/m^2) nitro-chalk

Polythene

Strawberries can be grown in a mulch of black or clear polythene to control weeds, conserve moisture in the soil and make strawing round the plants unnecessary. They can only be grown as single-spaced plants and not as a matted row; the stolons should be systematically cut off. Black polythene gives a complete control of weeds but delays ripening because under it the soil is colder. Clear polythene allows weeds to flourish under it unless herbicide is applied before it is put down, but slightly advances fruit ripening.

Polythene of 500 gauge should be strong enough to last for three or four years, whilst 50 gauge polythene rarely lasts for more than one season. The value of polythene in conserving moisture is equivalent to 2 in (5 cm) of rain.

One method is to plant the strawberries and lay the polythene on top of them, after which holes should be cut in the polythene with a knife, above the plants and the leaves pulled through the holes. Alternatively, the polythene can first be laid on the soil and the runners planted through holes made in the polythene. In this case, the holes should be made with either a blow torch or a knife.

The polythene sheet should be 24 in (60 cm) or 30 in (75 cm) wide. The sheets should be secured by burying 3 in (7.5 cm) of the edges in the soil. The soil between the strips of polythene should receive immediately an application of Weedex at one of the rates recommended on page 120, but none should fall on the polythene from where it might be washed by rain on to the strawberry plants. Alternatively, weeds that grow on this strip of soil should be burned off by applications of Weedol applied with a dribble bar, as long as none is allowed to fall on the strawberry leaves.

De-blossoming

Flowers will appear in the autumn on summer planted cold-stored runners and if the following summer's crop is to be a maximum one, these flowers should be removed. Freshly dug plants and pot grown plants bear their flowers in the month of

May after planting. Provided the plants are growing strongly, that is to say producing large new leaves, these flowers should normally be left in place to set and ripen fruit. However, if growth is poor for any reason, such as very late or poor planting or pest damage, the flowers should be cut off as soon as the length of the truss stalks allow this to be done. This enables the plants to devote all their resources to overcoming the previous check to their growth and building up strong crowns or producing new runners, so that a heavy yield of fruit should be produced in the first full cropping year.

'Perpetual' fruiting varieties should not be allowed to produce fruit at the same time as the heavy yielding summer varieties, but should be cropped later in August and September. The time that usually elapses between flowering and fruit ripening should be 42 days. The flowers should be cut off 'perpetual' fruiting plants until the last week of May or the first week of June if they are to produce the maximum amount of ripe fruit at the end of July or beginning of August, after the other varieties have finished cropping. De-blossoming 'perpetual' varieties increases the amount of fruit that will be picked during the autumn, but the amount will be less if they are not de-blossomed and allowed to crop during the summer as well.

Strawing

The purpose of strawing is to keep the fruit clean, and to provide a clean surface on which to place picking containers and on which to walk. Straw does increase the possibility of frost damage; therefore, strawberries that are not grown through polythene sheeting should be strawed at the beginning of June when the risk of spring radiation frosts is over. Barley straw is to be preferred as it is soft and pliable and easy to apply. One rectangular bale should be sufficient to straw a bed of 48 sq yds (40 m^2). The straw should be tucked round single plants and under the fruiting trusses; with the matted row system it is only necessary to cover the pathways and tuck the straw under the trusses that are growing into the rows. Strawberry mats can only be used on single plants, are too expensive and are never made large enough to protect all the fruit. If straw is not available, it is easy to make polythene mats of the correct size out of scrap polythene, used compost bags or fertiliser bags. Pierce the polythene with small holes so that pools of water do not lie round the plants.

Treatment of the Runners

Newly planted strawberry runners that are growing well should start throwing out stolons and new runners from mid-June onwards. Where the bed is to be grown as single-spaced plants, the stolons should be cut off as a matter of routine every seven to ten days. This concentrates all the energies of the plants into forming strong central crowns. Where matted rows are to be formed, the stolons should be allowed to grow and the first seven to ten runners from each parent plant should be allowed to root in the band of soil 8-10 in (20-25 cm) on either side of the rows of parent plants. Ideally the runners should be evenly spaced out over this band of soil and kept in place by stones or wire pins placed over the stolons. Any further stolons that come from the parent plants, or stolons from the rooting runners, should be systematically removed.

A less efficient but more usual method is to roughly pull the stolons and runners into the 8-10 in (20-25 cm) band. When it is estimated that the required number of runners have been trained in, new stolons and their unrooted runners that grow into the pathway should be cut off at regular intervals of time with a sharp spade or turf-edging iron.

One method which is not recommended is to allow the stolons to grow and runners to be left untended so that by October or November all the soil will be covered with rooted runners. Those stolons and runners that cover the pathways, which are usually 12-16 in (30-40 cm) wide, should be destroyed either by forking out and removing from the bed, or by destruction with one or two carefully applied applications of Weedol, or, as a last resort, by digging them into the soil.

The highest yield invariably comes from well grown matted rows which should yield 40 per cent more fruit than spaced single plants.

Watering

Strawberries, more than any other fruit crop, respond to watering in times of drought. When drought conditions occur, which usually means that no rain has fallen over a period of three or four weeks, strawberries should be watered with a garden sprinkler if this is possible. On normal soils of good depth and texture, this should not be necessary at the earliest before the fruit begins to swell at the end of May or beginning of June. If strawberries are watered before this, the leaves grow too vigorously and at the expense of fruit production. The exception to this is where strawberries are grown on shallow, sandy or gravelly soils; in these situations the beds should be watered on any occasion when drought conditions cause wilting of the leaves. A rough guide to watering would be to apply every week, 4.5 gallons per sq yd (20 l/m^2) of water to light soils and 9 gallons (40 l) of water every 14 days to heavier soils, (1 in equals 4.5 gallons per sq yd) throughout the period of the drought. A good way of measuring the amount of water applied to a bed would be to distribute two or three straight-sided tins round the bed under the sprinkler and continue watering until the average depth of water in the tins is 1 or 2 in (4.5 or 9 gallons per sq yd). Applying less than 1 in (2.5 cm) is a waste of time as too large a proportion evaporates and surface rooting is encouraged.

Cold-stored runners should always be watered within 24 hours of planting. Freshly dug runners, those in Jiffy 7s or out of pots planted during the summer period should be watered within two to three days of planting. Apply 4½ gallons of water per sq yd (25 l/m^2) so that the soil is wetted to a depth of 12 in (30 cm).

Picking

Strawberries are ready for picking when all the surface of the fruit has attained the shade of red that is characteristic of the variety. 'Cambridge Favourite' will be a paler red than the darker coloured 'Tamella', 'Domanil' or 'Totem' varieties.

At this stage they will have developed their full flavour, sweetness and aroma. Also by then, their keeping quality is little more than overnight at room temperature or 24 to 48 hours when kept in a refrigerator just above freezing. Strawberries are best picked like this for eating fresh, freezing or making into jam. For bottling or canning, the berries should be picked as soon as the skin has turned red, that is to say when they are slightly under-ripe.

It is also possible to keep strawberries for seven to ten days after picking if required. All berries which have turned two-thirds pink in colour whilst the other third is still white, should be picked unbruised, complete with their calyces. They should be stored in a refrigerator at 35°F (2°C) where they will gradually ripen and remain in a satisfactory condition for up to ten days; however, the flavour will not be equal to that of berries ripening naturally on the plants.

Strawberries should always be picked with the green calyx in place and without bruising the flesh. The stalk of each berry should be taken between thumb and forefinger and the stalk severed by pressing the thumbnail against the forefinger. A little practice enables this to be done so that the berries can be placed in the picking container without the fruit being touched or bruised. Fruit should always be picked into shallow containers with the berries lying not more than 2 in (5 cm) deep. Piled deeper than this, the weight of fruit above will bruise the lower layers of fruit.

Cleaning Up after Picking

The flower buds form inside the crowns of strawberry plants during August and September; therefore, it is important that the plants should be allowed to function with maximum efficiency at this time. It may not be generally realised that strawberry plants produce two sets of leaves. The first set grows away in the spring and nourishes the fruit, after which it serves no useful purpose and is probably a drain on the plant, particularly if the roots are not well supplied with water. The second set begins growing immediately after picking is finished and provides carbohydrate for the initiation of flower trusses within the crowns that will give rise to the following year's fruit.

Immediately (within two or three days) after picking has finished, the old leaves should be removed so that the new leaves are exposed to the light and can work at maximum efficiency. In the north of England and in Scotland this invariably leads to a significant increase in yield, presumably because the new leaves do not receive sufficient light when they are shaded by the old leaves. The leaves should be removed by cutting the leaf stalks as short as possible just above the crowns with a pair of hedging shears. Any stolons that have grown out should also be cut off. If weather conditions at the time allow, burning off is a good alternative to cutting, provided it can be done correctly without causing a nuisance to the neighbours and the strawberries are not growing through a polythene mulch, which would also be destroyed by the fire. The correct procedure should be to tease the straw with a fork so that it lies on top of the plants. On a dry day when the straw is bone dry and a moderate wind is blowing down the rows, light the straw on the windward side of the bed, so that there is a fast moving fire which will consume the leaves without harming the crowns. An advantage of burning is that it destroys any red spider mites, tarsonemid mites, greenfly, mildew and any other troubles that may have been present on the old leaves. Even so, whichever method of cleaning up the old strawberry bed is used, the new leaves should be kept under careful observation and if any of these troubles are found, the necessary protection measures should be applied.

It is sometimes recommended that applications of organic or artificial fertiliser should be made to strawberries after harvest. These are generally a complete waste, as the nutritional requirements are best met in late winter when the soil is full of moisture. The exception to this could be if the bed was lacking in vigour during the previous summer months; it could then be beneficial to broadcast:

$\frac{1}{3}$-$\frac{1}{2}$oz per sq yd (12-18 g/m^2) nitro-chalk

The bed should be thoroughly weeded and an application of herbicide made as recommended on page 120. Any stolons and runners that grow during the autumn should systematically be cut off as they appear, or destroyed with an application of Weedol.

Production of Early Strawberries

Strawberries can be produced early in the season from mid-May onwards in the south of England and the end of May in Scotland. Plants grown in frames, polythene greenhouses and glass cloches commence ripening their fruit approximately four weeks before the outdoor crop. Those grown under low polythene tunnels will only be two weeks earlier. The variety 'Redgauntlet', when grown in this way, is perpetually flowering and, in the south of England, will bear fruit throughout August, September and October. The basic cultural practices used for outdoor production apply equally well to early strawberry production. Cloches and frames are used less frequently than they used to be because glass is expensive, easily broken and difficult to handle. Glass frames produce earlier fruit than low polythene tunnels because the glass retains sun heat more effectively and is more efficient in preventing frost damage to the flowers.

Frames of any shape or size may be used provided they have at least 9 in (22.5 cm) of headroom and a satisfactory depth of reasonably fertile soil. The broken glass in frames can be replaced by polythene.

All sizes of glass cloche may be used, provided the plants under them are spaced correctly. The most suitably sized cloche measures 24 × 24 in (60 × 60 cm) with 9 in (22.5 cm) sides.

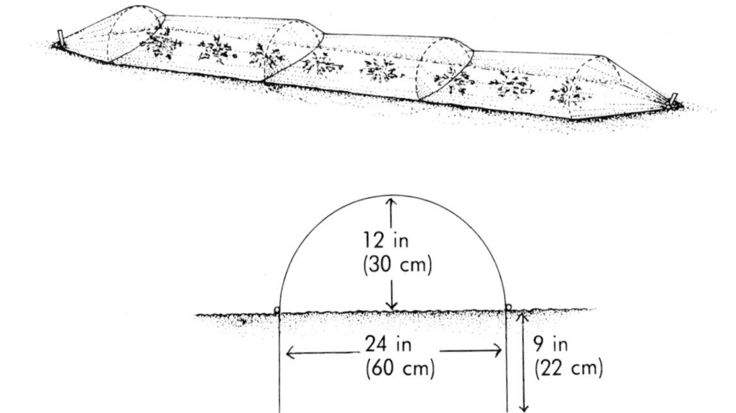

Figure 9.1: Low Polythene Tunnel Cover for Strawberries

Figure 9.2: Shape and Dimensions of Wire Tunnel Hoop

Low polythene tunnels should be constructed with hoops made from 6 gauge (5 mm) galvanised wire, spaced 48 in (1.2 m) apart, on top of which is tied 48 in (1.2 m) wide 50 gauge clear polythene. The width of soil covered by the tunnel should be 24 in (60 cm) wide (see Figure 9.1).

Standard high polythene tunnels are 14 ft (4.3 m) wide, 80 ft (24 m) long and covered with 600 gauge ultra-violet-light inhibited polythene supported on hoops of 15 mm galvanised steel tubing. Smaller sized tunnels in width and length than this can be constructed, provided UVL polythene film can be purchased economically.

Low polythene tunnels provide the cheapest method of producing fruit 14 days earlier than the outdoor crop. By burying the edges of the polythene in the soil, ripening can be brought forward by another seven days. The 6 gauge (5 mm) galvanised wire should be cut into 65 in (1.63 m) lengths and bent into a semi-circle 12 in (30 cm) in radius. Nine inches (22 cm) from the lower ends of the semi-circle, the wire should be bent round a ⅓ in (5 mm) diameter tube in order to form a small loop at the top of the legs of the hoop (Figure 9.2).

Planting

The soil should be prepared, manured, fertilised and Covershield applied as for open-ground plants, though nitrogen should be used sparingly as an excessive amount encourages too much leafy growth which in turn retards ripening. Plants should be of good quality with large crowns and good root systems. There is a choice of the following types of plant:

Cold-stored Runners. Planted during July.

Plants in Jiffy 7s or Pots. Planted during July/August.

Open Ground Runners. Planted towards the end of August and beginning of September. These should be raised by the gardener from Certified plants planted the previous autumn – a selected and restricted number should be allowed to root and surplus runners should be cut off.

Varieties

'Cambridge Favourite', 'Redgauntlet' and 'Tamella' usually produce heavier yields than other varieties. 'Cambridge Vigour', 'Gorella' and 'Pantagruella' will be earlier but lower yielding. Any variety can be grown but it is wise to select early-ripening ones and avoid those that are very vigorous with a lot of foliage.

Planting Distances

The distance between plants should be varied according to soil fertility, type and quality of plant and the district where they are grown. As the growing period between planting and winter setting in will be longer in the south and shorter in the north (UK), planting distances should be wider for the former and closer for the latter. In Scotland, the range of distances for fresh dug runners under tunnels and cloches 24 in (60 cm) wide should be 9-10 in (22.5-25 cm) between two rows and 9-10 in (22.5-25 cm) apart, staggered in the row. Planting distances in frames and polythene tunnels should be similar but pathways are required for tending three or four rows either side. In the south of England at the other extreme, cold-stored runners or Jiffy 7s should be planted under cloches or low polythene tunnels 10-12 in (25-30 cm) apart in a single row and in frames or high tunnels 10-12 in (25-30 cm) between the rows and 10-12 in (25-30 cm) apart staggered in the row. Where more than one row of cloches or low polythene tunnels are established, the minimum distance from centre to centre should be 36 in (90 cm) which leaves a 12 in (30 cm) path for walking. In July, the tunnels or cloches should be removed and the rows can be cropped in the years following as an ordinary strawberry bed.

Covering

The time for covering plants which have been grown in the open should be either the last week of February or the first week of March (UK). Covering earlier than this gives rise to too leafy a growth and the berries do not ripen any earlier. There are three methods of covering: glass cloches, low polythene tunnels and high tunnels.

Glass cloches should be placed over the plants so that the gaps between the cloches are as small as possible, and the ends of the cloches should be closed with sheets of glass or board.

With low polythene tunnels, the correct length of polythene sheet should be stretched over the hoops which should be 48 in (1.2 m) apart, and the ends of the polythene bunched together and tied to a stake driven into the soil. The polythene should then be pulled down over each hoop and a piece of polythene twine tied from loop to loop to prevent any movement of the polythene. Pressing the hoops 1-2 in (2.5-5 cm) into the soil and burying the edges of the polythene will help raise the temperature under the polythene and forward the time of ripening (see Figure 9.1).

In the first year, high tunnels are covered with polythene sheet at the end of February or the first week of March (see also p.

128). In the second year, the polythene sheet is already in place but the doors at both ends should remain fully open from August to February.

Growing Instructions

Before the plants are covered, all dead and withered leaves should be removed, the soil weeded and hoed and on all but the most fertile soils, ⅓ oz per sq yd (12 g/m^2) of nitro-chalk should be carefully sprinkled on the soil round the plants. A second application of Covershield should be made. The undersides of the leaves and the unopened leaflets should be carefully examined for the presence of red spider mites and aphids. If any are found, the plants should be sprayed with insecticide (see pp. 136-8). It would be a wise precaution in any case to apply insecticide to control these pests and to add a fungicide to control mildew.

The strawberries require little attention for the following six to eight weeks though high tunnels and frames should be ventilated if the temperature rises above 65°F (18°C).

The plants should come into flower towards the end of April and the berries ripen four to five weeks after that. Temperatures can rise to too high a level under all forms of cover and on sunny days the doors of high tunnels should be fully opened, frame lights raised, every fifth cloche removed and the polythene of low tunnels pushed up at every fifth hoop. Opening up the cover is particularly important during the flowering period to allow bees and other insects to pollinate the flowers. If a large area of strawberries are grown in this way, it might be well worthwhile borrowing a hive of bees to make sure that the flowers are pollinated.

Before and after flowering the leaves should be carefully examined for the presence of pests and diseases, and spraying carried out when necessary. It is advisable as a matter of routine that botrytis fruit rot fungicides should be applied at first flower, full flower and 14 days later. The leaves of protected strawberries are very tender and can be damaged by chemical sprays; therefore pesticides should be mixed accurately and applied on cloudy days. To prevent the berries lying on the soil and becoming dirty, straw or polythene should be put round the plants. Although some of the rain that falls will sink down to the roots it is rarely, if ever, sufficient for the needs of the plants and they should be watered if they wilt before the flowers set their fruit. The equivalent of 2 in (5 cm) of rain should be applied when the first berries turn white and this should be sufficient for the ripening period. On many sunny days during May and June the temperature round the plants will become too hot and ventilation similar to that during flowering should be given. All openings should be covered with netting to keep birds away from the berries.

Autumn Cropping 'Redgauntlet'

The variety 'Redgauntlet', when grown in this way in southern parts of the UK, bears a second crop of flowers in July and August and ripe fruit throughout August, September and October. Immediately after picking has stopped in June the covers should be removed, all the old leaves cut off and any pests or diseases present controlled by spraying with the appropriate chemical. The soil should be given a thorough soaking with water – the equivalent of 2 in (5 cm) (9 gallons per sq yd 40 l/m^2) of rain and if drought conditions occur, further water should be applied. The plants should be re-covered during mid-August and before mid-September at the latest, when fruit ripening is very slow and the berries become very susceptible to botrytis grey mould infection.

Production in Barrels

Growing strawberries in barrels can only be recommended where open ground is not available or adverse soil conditions make it impossible to grow the crop. The method is expensive, time consuming and skill is required for watering the plants correctly. Strawberries have been grown in second-hand wooden barrels for many years but although these barrels are long lasting, they are difficult to obtain and are costly. Instead, it is possible to buy a number of different makes and sizes of plastic barrels specially made for strawberry production – from garden centres and through mail order companies – which accommodate 30-50 plants in a very small space. Those in the cheaper range are not likely to have a useful life of more than four or five years.

Plastic barrels are usually 24 in (60 cm) tall and 18 in (45 cm) in circumference. The planting holes are 2 in (5 cm) in diameter and spaced 6 in (15 cm) apart with a staggered configuration (see Figure 9.3). A wooden barrel can be made into a strawberry barrel by removing the top, and boring holes of 2 in (5 cm) diameter in the sides with a brace and bit. The holes should be spaced 6-7 in (15-17.5 cm) apart, also in a staggered configuration. Six ½ in (1.25 cm) holes should be bored in the base of the barrel, so that excess water can drain out of the compost.

Figure 9.3: Plastic Strawberry Barrel

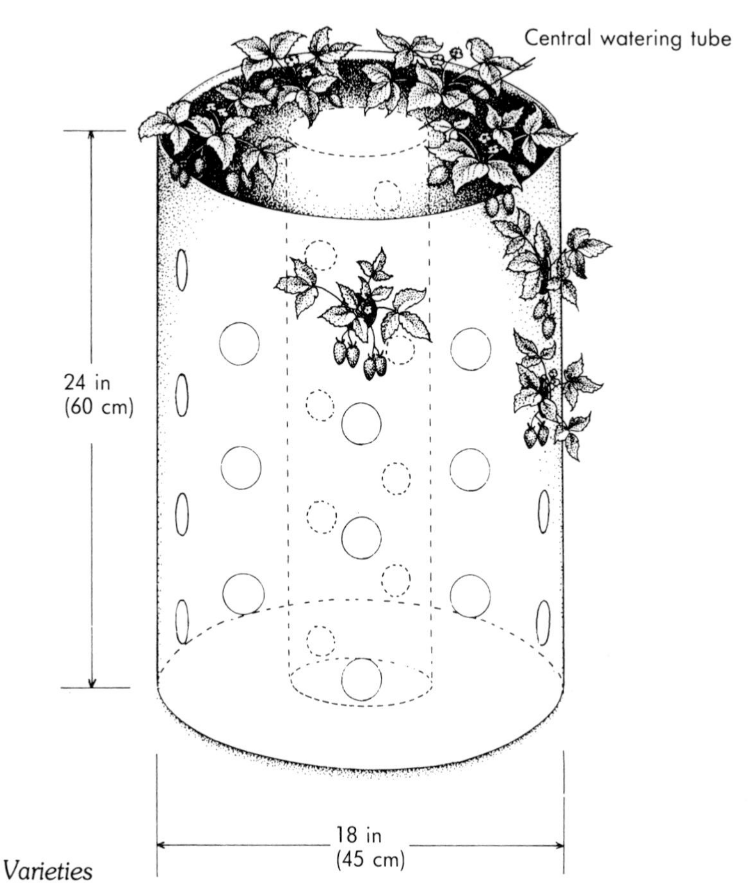

Varieties

To obtain the best return for the time and trouble involved, it is considered that heavy-yielding varieties such as 'Cambridge Favourite', 'Tamella' or 'Hapil' should be planted. There is no reason why 'perpetual' fruiting varieties should not be planted, though their yield may be less than that of summer fruiting ones. Each plant should yield 4-6 oz (112-168 g) of fruit in the year after planting and two to three times this amount in the years following.

Compost

The ideal compost is one that retains water and nutrient for a long period but does not become waterlogged when over-watered. The most suitable medium for growing strawberries in barrels would be:

> Three parts of a peat potting compost and one part of ⅛ in (0.3 cm) limefree grit.

The composition of composted garden waste is so variable that it should not be used alone or in a mixture with other materials for filling a barrel.

Drainage

Problems with over-watering and under-watering the top and bottom of the compost respectively are likely to be encountered. These difficulties should be overcome by making a watering and drainage channel down through the centre of the compost. Proprietary barrels are usually supplied with a plastic channel. Alternatively a channel can be made from 4 in (10 cm) diameter plastic drainpipe. Provided 1 in (2.5 cm) holes are drilled in the pipe at distances of 4 in (10 cm), the pipe may be left in position. Alternatively the drainpipe, or a channel formed from three 5 in (12.5 cm) wide pieces of wood, could be placed in the barrel whilst filling with compost takes place. The channel should be filled with crocks or rubble after which the pipe or wood should be withdrawn.

Filling the Tub and Planting

If the compost is dry, it should be thoroughly moistened with water to make filling and planting easier.

Fill the barrel with compost to a level of 2 in (5 cm) above the lowest ring of holes. Consolidate the compost, particularly near the sides, using a piece of wood to ram down the mixture. The level of the compost should settle just below the tops of the holes. If it does not, add or take away the necessary amount of compost.

The leaves of each strawberry plant, whether one with bare roots or growing in a Jiffy 7 (see page 117) should be inserted from the inside through the hole so that the base of each crown is in line with the side of the barrel at the top of the hole and the roots spread out on the compost.

It is a great help if each strawberry plant is pushed through a small slit approximately 1 in (2.5 cm) long in the centre of a piece of woven plastic measuring 4 in × 4 in (10 cm × 10 cm) which is placed inside the barrel and over the planting hole. The plastic prevents the compost escaping through the holes at the time of planting and when watering takes place.

Further layers of compost should be added and compressed so that all the holes can be planted and the tub finally filled and consolidated with compost to within 1 in (2.5 cm) of the rim. Strawberry plants should also be planted on the top of the barrel.

A test of the firmness of planting is that, if an attempt is made to pull a plant out of the tub by one of its leaves, the leaf tears before the plant moves. Do not apply this test to plants in Jiffy 7s. The nylon net round the root-ball of plants in Jiffy 7s should not be removed before planting.

Watering

Composts are designed to hold as much water and nutrients as possible and prevent the ill effects of over-watering, that can lead as often as not to the death of plants. Too frequent watering will also remove nutrients from the compost.

No hard and fast instructions can be given for watering because plants require smaller amounts of water when they are small, when temperatures are low and under dull light conditions. Frequent watering is required when temperatures are high, the sun is shining the plants are large and fruits are swelling.

After planting the central tube should be filled with water so that all the compost is thoroughly wetted; probably a further watering will not be required for at least a month, and with autumn planting, the next watering could be in the following spring. Then, watering frequency could be every three weeks, reducing to once a week at flowering time and as close as every two or three days when the berries are ripening and temperatures are high. That watering is not done frequently enough will be indicated by leaves on one or two plants beginning to wilt. Another serious cause of wilting could be that the roots are being eaten by the grubs of the vine weevil.

Feeding

Like over-watering, application of too much nitrogenous fertiliser adversely affects the cropping. Strawberries will crop best if they have the appearance of suffering from a slight deficiency of nitrogen and the leaves have a slight yellowish-green colour. In the first year of growing strawberries in tubs the nutrients in the compost should be sufficient to provide satisfactory growth until the plants come into flower. Only after flowering, and also in the autumn, is feeding likely to be considered necessary. The plants should not be fed again in the following spring unless the leaves appear pale green. A rather too yellow-green leaf colour is an indication that the plants need feeding. To one of the weekly waterings add one teaspoonful of Phostrogen to each gallon (4.5 l) of water that is poured into the central tube.

Plant Care

The best position for a tub is in full sunlight, but sheltered from the prevailing winds.

Tubs that have been planted in late summer or early spring and have strongly growing plants in them should be allowed to flower and fruit. If, for any reason, the plants are growing weakly, the flowers should be cut off to enable the strength of the crowns to be built up for the next crop.

As production of strawberries in tubs is expensive and time-consuming, it is important to ensure that all the flowers set their fruits. It is therefore recommended that when grown on a patio, or in a greenhouse, the flowers should be pollinated with a fine brush. This involves lightly passing the brush over the newly opened flowers so that pollen is transferred from the anthers to the stigmas.

From June onwards the plants will start producing stolons on which runners will form. These stolons should be cut off as soon as they appear with the exception of the autumn fruiting varieties, which will bear fruit on them.

The strawberry produces two sets of leaves each year – a spring set that supports the fruit and an autumn set which builds up new crowns for the following year's crop. Immediately after the plants have stopped fruiting, all the leaves should be cut off as close to the crown as possible so that the new leaves are fully exposed to the light as soon as they appear. In late winter the set of leaves that carried the plants through the autumn should be cut off. The leaves of 'perpetual' fruiting varieties should not be cut off in midsummer.

Annual Cropping

The original plants should crop for two or three years if they are carefully looked after. In the second and third year, success will depend upon maintaining the nutrients in the compost at a satisfactory level. For this purpose liquid feed at the rate of one level teaspoon of Phostrogen to the gallon (4.5 l) at each watering, to maintain a moderate fresh green colour and size of leaf. If the leaves turn dark green and grow too large, the addition of feed to the water should stop.

Pests and Diseases

It is essential to spray strawberries every year to control botrytis grey mould which, in wet picking seasons, can be responsible for the rotting of all the berries. Varieties that are known to be susceptible to powdery mildew should have routine sprays of fungicide applied to them. The leaves of strawberry plants are always liable to be infested with aphids (greenfly) and red spider mites, and in particular a close watch should be kept for the appearance of these pests during the period before flowering, so

Table 9.1: Strawberry Protection Chart

Problem	Chemical	Key	Product	April	May	June	July	Aug.	Sept.	Oct.
Aphids	Dimethoate	1	Boots Greenfly & Blackfly Killer							
		6	Murphy Systemic Insecticide							
	Fenitrothion	7	PBI Fenitrothion							
	Malathion	6	Murphy Liquid Malathion							
		7	Malathion Greenfly Killer							
	Permethrin	7	Bio Sprayday							
		2	Fisons Whitefly & Caterpillar Killer							
		3	Picket							
	Pirimiphos-methyl	3	Sybol 2							
Botrytis Grey Mould	Benomyl	3	Benlate plus Activex							
	Carbendazim	1	Boots Garden Fungicide							
	Thiophanate-methyl	6	Murphy Systemic Fungicide							
Chafer Beetle	Phoxim	2	Fisons Soil Pest Killer							
	Pirimiphos-methyl	3	Sybol 2							
Clay Coloured Weevil	Fenitrothion	7	PBI Fenitrothion							
	Pirimiphos-methyl	3	Sybol 2							
	Carbaryl	1	Boots Garden Insect Powder							
Green Capsid Bug	Malathion	7	Malathion Greenfly Killer							
		6	Murphy Liquid Malathion							
	Fenitrothion	7	PBI Fenitrothion							
	Pemethrin	7	Bio Sprayday							
		2	Fisons Whitefly & Caterpillar Killer							

Pest/Disease	Active ingredient		Product
		3	Picket
	Pirimiphos-methyl	3	Sybol 2
Leatherjackets	Diazinon	6	Root Guard
	Phoxim	2	Fisons Soil Pest Killer
	Pirimiphos-methyl	3	Sybol 2
Powdery Mildew	Benomyl	3	Benlate plus Activex
	Carbendazim	1	Boots Garden Fungicide
	Bupirimate and Triforine	3	Nimrod T
	Sulphur	5	Comac Cutonic Sulphur Flowable
Red Spider Mite	Malathion	7	Malathion Greenfly Killer
		6	Murphy Liquid Malathion
	Pirimiphos-methyl	3	Sybol 2
	Dimethoate	1	Boots Greenfly & Blackfly Killer
		6	Murphy Systemic Insecticide
Slugs and Snails	Methiocarb	7	Slug Guard
Strawberry Blossom Weevil	Fenitrothion	7	PBI Fenitrothion
	Malathion	7	Malathion Greenfly Killer
		6	Murphy Liquid Malathion
Strawberry Ground Beetles	Methiocarb	7	Slug Guard
Strawberry Rhynchites	Fenitrothion	7	PBI Fenitrothion
	Malathion	7	Malathion Greenfly Killer
		6	Murphy Liquid Malathion
Tortrix Moth Caterpillars	Fenitrothion	7	PBI Fenitrothion
	Malathion	7	Malathion Greenfly Killer
		6	Murphy Liquid Malathion
	Permethrin	3	Picket

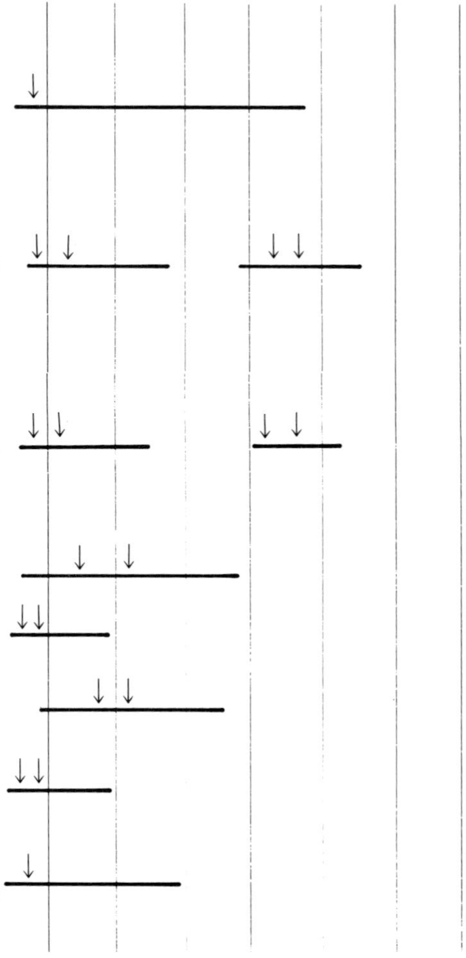

Problem	Chemical	Key	Product	April May June July Aug. Sept. Oct.
Vine Weevil	Pirimiphos-methyl	3	Sybol 2	
Verticillium Wilt	Benomyl	3	Benlate plus Activex	
	Thiophanate-methyl	4	Fungus Fighter	
		6	Murphy's Systemic Fungicide	
Wireworm	Diazinon	6	Root Guard	
	Phoxim	2	Fisons Soil Pest Killer	
	Pirimiphos-methyl	3	Sybol 2	

Key: ——— = Infestation periods ↓ = Timing of pesticide application
× Summer fruiting varieties.
×× Autumn fruiting varieties.

1 The Boots Company PLC, Nottingham.
2 Fisons PLC, Paper Mill Lane, Ipswich.
3 ICI Products Woolmead House, Farnham, Surrey.
4 May & Baker, Regent House, Hubert Road, Brentwood, Essex.
5 McKecknie Chemicals Ltd, PO Box 4, Widnes, Cheshire.
6 Murphy Chemical Co., Latchmore Court, Brand Street Hitchen, Herts.
7 Pan Britannica Industries, Waltham Cross, Herts.

that when necessary insecticides can be applied then to avoid the possibility of infestations occurring later, during the flowering period, and to avoid killing bees and other pollinating insects.

Propagation

Attention has already been drawn to the inadvisability of propagating strawberries indiscriminately because in most situations they become quickly infected with virus diseases. However, there are three methods of propagation. All varieties produce runners at the ends of their stolons and these root into the soil; in approximately six weeks they will have grown sufficient roots to enable them to be dug up and transplanted successfully.

The second method of propagation is to cut off the plantlets at the ends of the stolons as soon as they have grown one or two new leaves and even before they have produced any roots. They should then be treated like any cutting and inserted into a peat/sand compost in a pot or tray and kept in a humid atmosphere in a greenhouse or cold frame.

The third method is to propagate the crowns of old fruiting plants. The whole plant should be dug up and the crowns then split up. Two inch (5 cm) lengths of crown should be selected and the two youngest leaves left in place. Placed in pots or trays of compost they will root like young plantlets. This method should only be used if the parent plant appears healthy.

Chapter 10
Diseases of Soft Fruit

Blackcurrant
Gooseberry

American Gooseberry Mildew (*Sphaerotheca mors-uvae*)

Mildew can be expected every year to infect and cause serious trouble to gooseberry varieties with the exception of those that are known to be immune to the disease.

The white mycelium makes its appearance evident in June on the undersides of the leaves and developing gooseberries. Later, the disease spreads to the upper surfaces of the leaves and to the stems. The mycelium is quite thick and turns brown when the resting spores of the fungus are formed. The mat of mycelium can be rubbed off gooseberries and they remain edible. Symptoms on the leaves and stems of blackcurrants are similar (see Plate 14) but infections can have a catastrophic effect on the growth of the bushes.

Control Measures New infections take place as early as April (UK) before the flowers open. It is important to prevent these infections early rather than wait until the berries are seen to become infected:

(1) The tips of any shoots that were infected during the previous summer should be cut off and burned.
(2) Avoid making excessive applications of nitrogenous fertiliser which give rise to soft growths that are susceptible to infection.
(3) Grow the variety 'Invicta', which is immune to infection. The blackcurrant 'Ben More' is also immune, whilst 'Ben Lomond', 'Ben Nevis' and 'Ben Sareck' are moderately resistant. Red and white currants, though not immune to mildew, are rarely troubled by the disease.
(4) Spray the bushes, making sure that all the leaves in the centres of the bushes are covered with fungicide:

    (a) before flower opening;
    (b) at fruit set;
    (c) 3 weeks later;

using:

	Available as
Bupirimate and Triforine	Nimrod T
or	
Sulphur	Comac Cutonic Sulphur Flowable

Blackberry Purple Blotch (*Septocyta ramealis*)

Blackberry
Hybridberry

The disease is usually seen first in midsummer when lengths of the new cane have turned purple in colour. The disease is usually worse near to the stool but all parts can become infected. In the fruiting year, growth of the infected canes is poor and buds in the infected areas may be killed.

Control Measures

(1) Train the new canes above or away from the fruiting canes.
(2) Spray:

(a) in May;
(b) in June;
(c) after harvest;

using:

	Available as
Copper	Comac Bordeaux Plus or Murphy Liquid Copper

When the bushes are sprayed with benomyl to control botrytis grey mould, control of purple blotch will be enhanced.

Botrytis Grey Mould (*Botrytis cinerea*)

All Fruits

The grey mould fungus attacks a large number of different kinds of plants and all the soft fruits. It is usually responsible for larger losses than any other disease, particularly on soft sappy growth when weather conditions are wet or humid. The disease can only infect dead or damaged tissues. It is important to appreciate that most fruit infection takes place during the flowering period and it is usually later in the summer during wet or humid weather that the disease develops. The fruits rot and are covered with a dense grey fungal mycelium, either whilst still hanging on the bushes (see Plate 15) or after they have been picked. As well as infecting fruits that have been injured by insects or other causes the fungus infects dying petals and stamens and can spread by contact to the developing fruits. The objective of spraying is to maintain a cover of fungicide over the flowers and developing fruits from flower-opening until shortly before picking.

Grey mould also infects and can kill the shoots, canes or branches of all the cane and bush fruits. This usually occurs when excessive applications of nitrogenous fertiliser have been made or the growth of bushes has been allowed to become overcrowded. Large branches of blackcurrants, gooseberries or redcurrants either fail to leaf out in the spring or suddenly collapse during the summer. Careful examination of such branches usually shows they had been damaged and invaded by the fungus.

Infections on raspberry canes can first be seen during July when the lower parts of the cane, usually between the nodes, turn bright purple in colour and the infections spread up and down the cane. During the winter the infected areas turn silvery-white and by the end of the winter black circular spots 1/10 in (2-3 mm) in diameter (fruiting bodies) can be seen embedded in the silvered areas. Buds situated within the silvered areas may fail to grow out or severe infections can kill the canes.

The new shoots of blackcurrants, gooseberries and redcurrants are also weakened or killed by the disease.

Control Measures Remove and burn or compost all dead plant debris lying about the fruit plot. Pick mouldy fruits and bury them.

Refrain from making excessive applications of nitrogenous fertiliser that give rise to over-vigorous sappy growth which is susceptible to injury and infection by the disease.

Do not allow excessive numbers of strawberry runners to root in matted rows. Raspberry canes should be thinned out to 8-10 per yd of row (10-12 per m) in May and June (UK). The centres of gooseberry and redcurrant bushes should be judiciously thinned out during June and the bushes can, with advantage, be summer pruned in late July or early August.

The maximum protection of the fruit can be obtained by spraying four times:

as the first flowers open;

at full flower;
14 days later;
14 days later.

As strawberries are most susceptible to and suffer greater losses from this disease, they should always be sprayed four times. In the drier parts of the United Kingdom it should only be necessary to apply the first two sprays to the other fruits. However, when spring frosts have occurred during the flowering season and probably injured the flower parts of blackcurrants and gooseberries, the full spray programme should be applied.

The only fungicides that are available to the amateur fruit-grower are of the systemic kind, whereby the fungicide is absorbed by the plant and translocated either through the leaf or to other parts of the plant. Unfortunately, strains of the fungus resistant to these fungicides may be present in some gardens so that control of the disease may not be as effective as formerly. The fungicides which are available are:

	Available as
Benomyl	Benlate plus Activex
or	
Carbendazim	Boots Garden Fungicide
or	
Thiophanate-methyl	Murphy Systemic Fungicide or Fungus Fighter

Cane Spot (*Elsinoë veneta*)

Blackberry
Hybridberry
Raspberry

The disease can first be seen from June onwards as small circular purple spots on the new canes. As the canes grow the spots enlarge and become eliptical in shape. The centres of the spots become sunken and silvery-white in colour whilst the edges remain purple. The infected tissues die and by the end of the following winter the dead and cracked underlying wood is exposed. In a severe infection the spots coalesce to form irregular cankers which may girdle and kill the canes. The leaves also become infected and can be covered with the characteristic silver and purple spots of the disease. Cane spot is worse in the high rainfall areas of the United Kingdom; where bushes have received excessive applications of fertiliser; and the canes are overcrowded.

Control Measures

(1) Do not apply more manure than is required to give a satisfactory length of cane.
(2) Hoe out in spring surplus raspberry canes that will not be required to bear fruit in the following year.
(3) Train the new canes of hybrid berries so that they will not be directly underneath the fruiting canes to prevent rain from washing spores of the disease from the old onto the new canes.
(4) Spray the new canes in mid May, early July and August to prevent infection by the disease, using:

	Available as
Copper	Comac Bordeaux Plus
	or
or	Murphy Liquid Copper
Benomyl	Benlate plus Activex
or	
Carbendazim	Boots Garden Fungicide

Blueberry

Canker (*Godronia cassandrae*)

Symptoms of this disease are that the leaves of individual shoots or branches suddenly wilt and die. If the stems on which these leaves are borne are examined, irregular brown-coloured areas surrounded by yellow to bright red margins will be found (see Plate 16). When these cankers girdle a stem or branch, the passage of sap to the upper portion is interrupted and the leaves wilt and die.

Control Measures As soon as a stem or branch dies, or the characteristically-coloured cankers are seen, cut the shoot or branch back below the canker into the healthy part of the stem or branch.

Blackberry
Hybridberry
Raspberry

Crown Gall (*Agrobacterium radiobacter var. tumefaciens*)

This is a bacterial disease that affects blackberries, hybrid berries and raspberries. It causes galls on the roots of woody convoluted tissue sometimes as large as tennis balls and longitudinal galls on the fruiting cane. Although severe infections appear horrific they

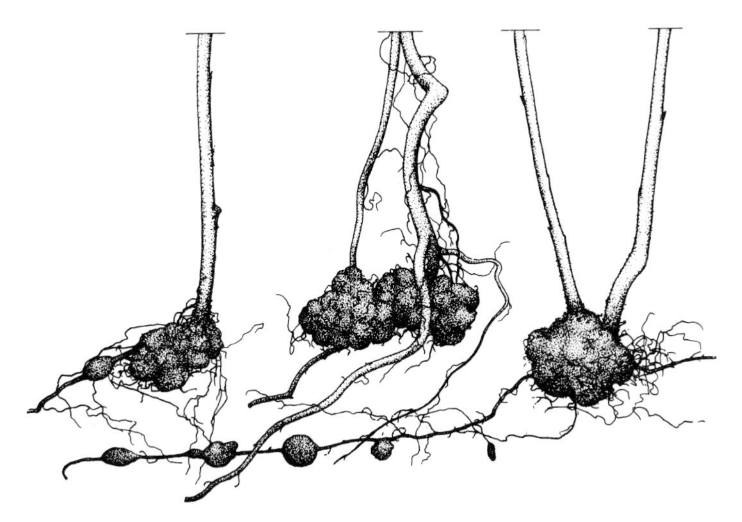

Figure 10.1: Raspberry Roots Infected with Crown Gall

may cause only minor reductions in yield. Infections may be severe one year and can almost disappear in the following year, and it is considered that many infections are related to frost and mechanical damage to the plants.

Control Measures

(1) Do not plant on poorly drained soils.
(2) Examine plants carefully for the presence of galls and return infected material to the nurseryman.
(3) Avoid damaging the canes. Adopting a training system where the new canes are tied in their fruiting position helps to avoid damage.
(4) Sprays containing copper applied for cane spot appear to reduce crown gall infection.

European Mildew (*Microsphaera grossulariae*)

Blackcurrant
Gooseberry
Redcurrant

This disease is of little importance as the thin, grey-coloured mycelium is usually confined to the upper surfaces of the leaves and rarely infects the fruit. A severe infection can cause a somewhat premature leaf fall.

Control Measures Fungicides applied for American gooseberry mildew will easily control this disease (see p. 140).

Crown Rot (*Phytophthora cactorum*)

Strawberry

Plants of any age may become infected by this disease, symptoms of which are seen in midsummer, or earlier under glass, when temperatures are high. First the younger leaves wilt, then the older leaves and the whole plant turns to a light green colour and dies. A closer examination will reveal symptoms of rot in the crown of the plant as well as the lower part of the stalks of the wilting leaves emanating from the infected crown. Infection comes from the soil but it can come in on purchased infected runners. 'Tamella' in particular is susceptible to infection by crown rot.

Control Measures There are no satisfactory ways of preventing infection of plants by this disease. Do not plant in soil on which plants were infected. Avoid planting susceptible varieties.

Blackcurrant
Gooseberry
Redcurrant

Leaf Spot (*Pseudopeziza ribis*)

A severe infection of leaf spot can completely defoliate currant and gooseberry bushes before the end of July. This not only reduces seriously the size of the berries and the weight of fruit but stops further growth of the new shoots and the number of flowers that will appear in the following spring.

Infection commences in late April on leaves at the base of bushes and increases rapidly under wet weather conditions. The first sign of the disease are a few angular brown spots accompanied by a yellowing of the surrounding surface of the leaves. Under favourable wet conditions to the fungus, the numbers of spots increase rapidly and coalesce until the leaves are killed and fall off the bushes.

Control Measures Leaf spot is fairly easily controlled by spraying with fungicide at:

(a) early flowering;
(b) fruit set;
(c) 3 weeks later.

In very wet summers a further application should be made after picking, using:

	Available as
Copper	Comac Bordeaux Plus
	or
or	Murphy Liquid Copper
Benomyl	Benlate plus Activex

or

Carbendazim	Boots Garden Fungicide

Powdery Mildew (*Sphaerotheca macularis*)

Strawberry

The disease causes serious reductions in plant vigour, crop yield and fruit quality. It appears on the plants from late April onwards, particularly during periods of dry weather. Small reddish-purple spots and patches occur on the upper surfaces of the leaves and on the lower corresponding surfaces. A close examination will reveal the presence of a white powdery mycelium. As the disease progresses, the leaves dry up and their edges curl upwards; infection of developing berries follows. Those that are infected early will be small and severely distorted whilst later infected ones have a dull appearance and prominent seeds.

Control Measures

(1) There is a considerable variation in susceptibility to mildew. 'Cambridge Favourite' is immune whilst 'Cambridge Vigour' and 'Gorella' are very susceptible to infection.

(2) Susceptible varieties should be sprayed at least three times:

(a) first flower opening;
(b) fruit set;
(c) 14 days later.

(3) Burning over the bed after picking reduces the possibility of infection during the autumn.

(4) Spraying with fungicide in the autumn should be carried out if the disease occurs, using:

	Available as
Benomyl	Benlate plus Activex
or	
Bupirimate and Triforine	Nimrod T
or	
Sulphur	Comac Cutonic Sulphur Flowable

Raspberry

Powdery mildew on raspberries is very similar to that on strawberries. The mycelium is present on the undersides of the leaves but causes a yellowing of the veins on their upper surfaces and this is very difficult to distinguish from virus infection. The effect on growth of the new canes is minimal and rarely, if ever, serious. The white mycelium growing on the surface of the berries detracts from their appearance, reduces size and can cause distortion.

Control Measures

(1) Fungicides applied for the control of botrytis grey mould can give a satisfactory control of mildew infection of the fruit.

(2) Where severe infection is not controlled by these fungicides spray in addition with Nimrod T at:

(a) full flower;
(b) 14 days later.

(3) Recommended fungicides are:

	Available as
Benomyl	Benlate plus Activex
or	
Carbendazim	Boots Garden Fungicide
or	
Bupirimate and Triforine	Nimrod T

Strawberry

Red core (*Phytophthora fragariae*)

Red core is a disease of the roots of strawberry, so called because, in late autumn or spring, when affected roots are split open longitudinally, the centres, instead of being white, are reddish-brown in colour. First indications of the presence of red core occur in dry weather during May or June when a number of plants will be seen to be growing less well than their neighbours. All the leaves will have a bluish-grey tinge and the outer leaves will turn yellow, orange and red in colour. The berries ripen prematurely, are small in size and negligible in weight if the infection is severe.

The disease is usually brought into a garden on infected uncertified runners. It is worst on heavy soils that are poorly drained but can also occur on lighter soils where a hard pan prevents water from draining away.

Control Measures

(1) Buy certified plants to reduce the risk of introducing the disease into the garden.
(2) Before planting, make sure that the soil is well drained by double digging to a depth of 18 in (45 cm).
(3) Heavy soils should have a tile or plastic drain installed so that surplus water is rapidly removed from the strawberry plot.
(4) The worst effects of the disease can be avoided on heavy soils by planting on 4-6 in (10-15 cm) high ridges.

Reversion Virus

Blackcurrant

Branches or whole bushes may suddenly cease to set their fruit. If other bushes are bearing full crops of blackcurrants, it is reasonable to assume that a bush (or bushes) is infected with reversion virus. This is confirmed by examining the leaves in June, which have a coarse appearance with a smaller number of veins than leaves on healthy bushes. Further confirmation of the presence of virus can be obtained by examining the flower trusses just before the flowers open. Flower buds of healthy bushes are hairy and this gives them a dull purplish-green colour, whilst the flower buds of infected bushes are almost hairless, somewhat shiny and purplish-red in colour.

Control Measures

(1) There is no cure for this disease, and infected bushes should be dug up and burned as soon as symptoms of the disease are diagnosed.
(2) Replant with certified bushes.
(3) As the disease is spread from infected bushes to healthy bushes by big bud mite, take the control measures used for this pest (see p. 155).

Rust

Blackcurrant (*Cronartium rubicola*)
Blackberry (*Phragmidium violaceum*)
Raspberry (*Phragmidium rubi-idaei*)

The rust diseases are bright orange-coloured spots to be found on the under surfaces of the leaves from late spring onwards. Later the spots turn dark brown. There are corresponding spots of paler colour on the upper sides of the leaves.

It is only in recent years that the disease has become serious on raspberries, causing premature leaf drop and reduction of vigour.

It is worst on 'Delight' and 'Glen Clova'.

Occasionally the disease causes the defoliation of blackcurrants. On blackberries the disease is similar to that on raspberries but the colour of the spots on the upper surfaces of the leaf is dark red or purple. The cut-leaved varieties are susceptible to rust.

Control Measures Pick off and burn infected leaves. Non-fruiting bushes can be sprayed with:

	Available as
Propiconazole	Murphy's Tumbleblite

Fruiting bushes should be sprayed with:

Copper	Comac Bordeaux Plus or Murphy Copper Fungicide

Blackberry
Hybrid Berry
Raspberry

Spur Blight (*Didymella aplanata*)

Spur blight is a disease that infects the new canes from the end of June onwards. It may prevent buds situated on infected areas from growing out, but the disease appears to be worse than it actually is, as usually, buds that are situated on infected parts of the cane are unaffected and grow normally. Symptoms of the disease are the appearance of purple patches surrounding the leaf bases, though intermediate pieces of cane between buds can be infected. The infections elongate and extend round the cane until whole lengths of cane become infected. Towards winter the purple colour turns silvery-white (see Plate 17) in which is situated large numbers of minute black spots which are sporing bodies and distinguish the disease from botrytis infection.

Control Measures

(1) Reduce applications of nitrogen when canes are growing too vigorously.

(2) In May and June, thin out the number of new canes to 8-10 per yd (10-12 per m) of row.

(3) The disease looks worse than it appears to be – if after measures (1) and (2) fail to reduce the incidence of the disease, when spraying for botrytis grey mould, spray the new canes as well with fungicides using:

	Available as
Benomyl	Benlate plus Activex

or

Carbendazim	Boots Garden Fungicide

Stamen Blight (*Hapalosphaeria deformans*)

Raspberry
Blackberry

Infected flowers are slightly larger than healthy ones, and the stamens are covered with masses of white spores which give the flowers a mildewed appearance. Berries that develop from infected flowers ripen earlier than healthy ones, are deformed and the drupelets ripen unevenly and are difficult to pick. Spores from the flowers infect the buds on the new canes.

Control Measures Pick off and burn infected flowers as soon as they are seen.

Verticillium Wilt (*Verticillium dahliae*)

Strawberry

This disease is more common in the southern parts of the United Kingdom and affects newly planted strawberries rather than fruiting beds. The plants are affected from June onwards; the outer leaves wilt and lie flat on the ground whilst undeveloped leaves are yellow and remain small. Black streaks are to be found on the leaf stalks and stolons, and if the crown is cut across, a dark brown discoloration of the conducting tissues is to be seen. Severely infected plants die, whilst mildly infected ones recover and crop fairly normally in the following year. The disease can be contracted from potatoes grown as the preceding crop.

Control Measures
(1) Avoid planting strawberries after potatoes.
(2) Do not take runners from an infected bed.
(3) Grow resistant varieties.
(4) If strawberries grown in the garden have been infected by this disease, drench newly planted runners with 10 fl oz (0.3 l) per plant after planting and again twice in the spring, four weeks apart using:

	Available as
Benomyl	Benlate plus Activex
or	
Thiophanate-methyl	Fungus Fighter or Murphy Systemic Fungicide

Chapter 11
Pests of Soft Fruit

All Fruits

Aphids (*Greenfly*)

All soft fruit bushes may become infested with aphids. They can cause serious physical damage to all the currants, gooseberries and strawberries and this makes it necessary, usually every year, to spray these crops with insecticide as a matter of routine. The damage aphids cause to raspberries, blackberries and hybrid berries is less serious and it is only in occasional years that infestations build up to such a level that preventive measures become necessary. Many of the virus diseases that infect soft fruit bushes are carried by aphids; spraying to kill aphids does not prevent infections taking place, but does reduce the spread of virus diseases within the fruit plot. Depending upon the insecticide used, several minutes or hours may pass before the aphids die and, during the period from when the aphid lands on a bush until it dies, infection of the bush may take place.

Blackcurrant
Redcurrant

Currant Sowthistle Aphis (*Hyperomyzus lactucae*) (Plate 18)
Currant Blister Aphis (*Cryptomyzus ribis*)

Two species of aphid cause severe damage to red and black currants. They hatch out from eggs laid on shoots during the previous autumn and feed on the underside of the leaves situated on the fruiting wood and later on the new shoots. When an infestation is severe the leaves fall off, fruit fails to set and the new shoots are stunted. The upper leaves are tightly rolled up and the lower ones covered with honeydew.

A third species is less serious as it does not cause leaf rolling, but lives on the undersides of the leaves and its feeding gives rise to the characteristic red blisters on the upper sides of the leaves.

Blueberry

Very occasionally aphids will be found feeding in the tips of new shoots.

Gooseberry Aphid (*Aphis grossularia*)

Gooseberry

The aphid causes similar damage on gooseberry bushes to that caused by leaf curling aphids on blackcurrants. If an infestation is allowed to go unchecked, its effect on the fruit and new shoots can be catastrophic.

Raspberry Aphid (*Aphis idaei*)
Raspberry Leaf Curling Aphid (*Amphoraphora idaei*)

Raspberry
Blackberry
Hybrid Berries

These fruits become infested with the raspberry aphid every year. They feed on the undersides of the leaves and in many years, apart from sucking the sap, appear to do little harm. Less common is the raspberry leaf curling aphid which, as its name indicates, causes a characteristic curling of the leaves at the tips of new shoots. These aphids appear to cause little harm and routine spraying for raspberry beetle with fenitrothion or malathion is sufficient to prevent damage. Occasionally, usually in drought years when they build up to such large numbers that the leaves become covered with honeydew, it will be necessary to make a special application of insecticide. Certain raspberry varieties – 'Autumn Bliss', 'Delight', 'Glen Moy', 'Glen Prosen', 'Joy', 'Leo' and 'Orion' – are resistant to colonisation by aphids.

Strawberry Aphid (*Chaetosiphon fragaefolii*)
Shallot Aphid (*Myzus ascallonicus*)

Strawberry

There are two kinds of aphid that commonly infest strawberries. The strawberry aphid is responsible for carrying virus from one strawberry plot to another and from plant to plant but is only occasionally present in such large numbers that it causes physical damage. It is very small and pale lemon in colour. The shallot aphid is greyish-green and much larger than the strawberry aphid. In April and May the numbers of shallot aphid can increase rapidly. The strawberry leaves curl up and the plants become very severely stunted. The aphids feed on the developing berries and, even if they ripen, they are severely distorted. At first only a few dwarfed plants will be seen but the infested area rapidly enlarges until the whole bed is affected.

Control Measures Winter. Blackcurrants, gooseberries and redcurrants should be sprayed with tar oil during the winter whilst they are dormant to kill the overwintering eggs. It has never been the practice to spray cane fruits at this time. Use:

	Available as
Tar oil	ICI Clean Up or Murphy Mortegg

Summer. The bushes and plants should be inspected regularly, particularly during April and May, so that measures can be taken before serious harm has been caused and in the case of some kinds of aphid, before the leaves have curled up, using:

	Available as
Malathion	Malathion Greenfly Killer or Murphy Liquid Malathion
or	
Fenitrothion	PBI Fenitrothion
or	
Dimethoate	Boots Greenfly and Blackfly Killer or Murphy Systemic Insecticide
or	
Permethrin	Bio Sprayday or Picket or Fisons Whitefly and Blackfly Killer
Pirimiphos-methyl	Sybol 2
Rotenone	Liquid Derris

Dimethoate should be kept for use when infestations are very severe and the leaves are curled up; otherwise spray with one of the other insecticides. Permethrin is a synthetically produced pyrethrum. Rotenone is a naturally-occurring insecticide that is extracted from a root – it is not very effective.

Big Bud Mite – see Blackcurrant Gall Mite.

Blackberry
Hybrid Berry

Blackberry Mite (*Acalitis essigi*)

The blackberry mite is more commonly called the redberry mite because, on blackberry, its feeding causes berry distortion and a number of the drupelets remain hard and red when the rest of the berry has ripened and turned black. The mites also infest loganberries and hybrid berries.

These minute transluscent mites that cannot be seen with the naked eye overwinter under the bud scales. They feed on the leaves before migrating to the newly opened flowers. There they feed near the calyx on the immature drupelets. This feeding is the cause of distortion and uneven ripening of the berries.

Control Measures There are no really effective methods of controlling this pest. The fungicides applied for botrytis grey mould probably somewhat reduce the numbers of mites. Spraying with sulphur from bud-burst to flower-opening at 14 day intervals might have a similar deterrent effect. If the infestation becomes so serious that few sound fruits develop, all the old and new fruiting cane should be cut down to ground level during the winter and burned although the crop will be sacrificed for one year. The sulphur should be applied as Comac Cutonic Sulphur Flowable.

Blackcurrant Gall Mite (*Cecidophyopsis ribis*)

Blackcurrant

The blackcurrant gall mite or big bud mite, as it is more usually called, is a microscopic insect. It emerges during April and May from infested buds and migrates to and feeds on the newly unfolding leaves until buds on the new shoots form. It enters these buds and continues multiplying and feeding until the following spring. The buds swell to a characteristic globular shape (see Plate 19) during the period August to December. Affected buds fail to grow and subsequently fail to flower. More seriously, the mites transmit reversion virus disease to the bushes; this virus prevents bushes from setting fruit.

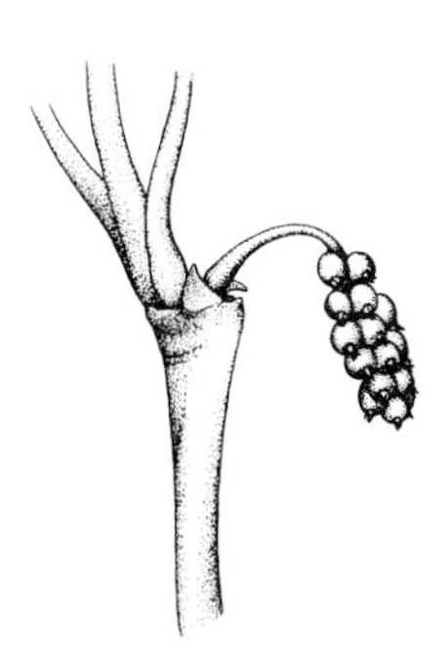

Figure 11.1: Blackcurrant Grape Stage, Development of Flower Buds

Control Measures Always buy certified bushes and never propagate from infested bushes.

Regularly inspect the bushes and, as soon as swollen buds are seen, pick off and burn them.

It is considered that bushes that have ten per cent of infested buds are unprofitable and should be dug up and burned.

There is no specific pesticide available which will control this mite. The risk of infestation can be reduced by spraying three times at ten day intervals from the 'grape' stage (see Figure 11.1) of flower development, using:

	Available as
Benomyl	Benlate plus Activex
or	
Carbendazim	Boots Garden Fungicide or Fungus Fighter
or	
Thiophanate-methyl	Murphy Systemic Fungicide

In addition, or as an alternative, spray twice before and once after flowering, using:

Sulphur	Comac Cutonic Sulphur Flowable

Grub out and burn any bush that is infected with reversion, whether or not it is infested with the gall mite.

Blackcurrant

Blackcurrant Leaf Curling Midge (*Dasineura tetensi*)

The midges lay eggs in the folded leaves of the new shoots during May and there are further generations throughout the summer. The eggs hatch out into small, white legless grubs and their feeding stops the leaves unfolding, which remain twisted and necrotic. A severe infestation can stunt the new growth.

Control Measures Thoroughly spray the tips of the shoots as soon as affected leaves appear and repeat the application 14 days later, using:

	Available as
Dimethoate	Boots Greenfly and Blackfly Killer or Murphy Systemic Insecticide

All Fruits

Chafer Beetle (*Melolontha melolontha*)

The grubs of the chafer beetle feed upon the roots of all plants but it is newly planted and established strawberries that they seriously damage, making the plants wilt and sometimes die. When fully grown, the grubs are 1½ in (4 cm) long, white in colour with a brown head.

Control Measures Where chafer grubs have given trouble in a garden, apply and work into the soil before planting:

	Available as
Phoxim	Fisons Soil Pest Killer
or	
Pirimiphos-methyl	Sybol 2 Dust

To protect newly planted and established strawberries apply as a drench ¼ pint (150 ml) of insecticide to each plant, using:

Pirimiphos-methyl	Sybol 2

Clay Coloured Weevil (*Otiorhynchus singularis*) **All Fruits**

This insect is rarely seen because it hides in the soil during the day and only comes out at night to feed. It feeds on the leaves and fruiting laterals, notching the main veins, leaf and flower stalks so that pieces of wilting and dead tissue are found hanging on the bushes. In a serious attack, a bush can be denuded of every piece of green leaf and stalk. The weevil can also kill or damage the shoots of newly planted raspberries and crowns of newly planted strawberries.

Control Measures Normally an attack starts at the beginning of May and as soon as symptoms are observed, spray or dust the lower parts of, and the soil at the base of, each bush. Two or three applications may be required to kill this pest, using:

	Available as
Fenitrothion	PBI Fenitrothion
or	
Pirimiphos-methyl	Sybol 2
or	
Carbaryl	Boots Garden Insect Powder
or	
Pirimiphos-methyl	Sybol 2 Dust

Currant Clearwing Moth (*Synanthedon salmachus*) **Blackcurrant Gooseberry Redcurrant**

This pest attacks blackcurrant and redcurrant and, less frequently, gooseberry bushes. The moth lays eggs on the shoots in June. The white caterpillars bore into and up the stems, either killing or making them liable to break off during the following summer.

Control Measures Where this pest is present the shoots should be tested each winter for the presence of the caterpillar by bending over each shoot to find out whether it will break off easily. Those that do so should be pruned back below the tunnel made by the caterpillar and burned.

Blackcurrant
Gooseberry
Redcurrant

Gooseberry Sawfly (*Nematus ribesii*)

The adult sawflies may commence laying eggs on leaves in the centres of the bushes as early as April. The caterpillars are green with black spots and heads and usually will not be noticed until after the fruit has set at the end of May, when they skeletonise the leaves. Second and third generations of caterpillar may appear in June and August. If infestations are not checked, the bushes can lose all their leaves and the berries do not swell. (See Plate 20.)

Currant bushes are sometimes attacked but rarely as severely as gooseberries.

Control Measures As soon as the caterpillars are seen the bushes should be sprayed, making sure that the leaves in the centres of the bushes are well covered, using:

	Available as
Rotenone	Derris Liquid
or	
Fenitrothion	PBI Fenitrothion
or	
Malathion	Malathion Greenfly Killer or Murphy Liquid Malathion

All Fruits

Green Capsid Bug (*Lygocoris pabulinus*)

This insect can infest and cause serious damage to the leaves and shoots of all the soft fruits but it only damages the fruits of strawberry. The capsid bug is a small pale green insect that feeds between the folded leaflets. In four to five weeks the insect reaches the adult stage when it is ⅓ in (8 mm) in length, bright green, and has developed wings and can fly. It can be distinguished from greenfly by the fact that it is very quick-moving and runs rapidly over the leaves.

The capsid bug feeds on the expanding leaves causing minute

brown spots which form conspicuous holes when the leaves are fully grown. The brown spots and holes are indicative of the presence of this pest. More serious is the killing of growing points causing branching and stunting of the shoots and canes. On strawberries, capsids feed on the developing fruits, giving rise to distorted flesh with patches of close-set seeds (cat-faced fruit).

Control Measures Capsids appear first on soft fruit bushes at the beginning of May. In June and July, adults can fly into the fruit plot from weeds such as nettles. As soon as the damage is observed, spray the bushes, making sure that growing points of shoots and the unfolding leaves of strawberry are covered with insecticide. Use:

	Available as
Fenitrothion	PBI Fenitrothion
or	
Malathion	Malathion Greenfly Killer or Murphy Liquid Malathion
or	
Permethrin	Fisons Whitefly and Caterpillar Killer or Bio Sprayday or Picket
or	
Pirimiphos-methyl	Sybol 2

Leatherjackets (*Tipula spp.*)

Raspberry
Strawberry

Leatherjackets, which are the larvae of crane flies or 'Daddy long legs', feed below soil level on the shoots of newly planted raspberry canes and the crowns of strawberry plants. They can devastate these crops.

The grubs, when fully grown, are 1 in (25 mm) long, legless and about the thickness of a pencil but, as they are the same colour as the soil, can be rather difficult to find.

Control Measures Fill a spraying machine or watering can with insecticidal solution and soak the soil round the neck of each plant with ⅕ pint (100 cc) of the liquid, using:

	Available as
Pirimiphos-methyl	Sybol 2

Alternatively, work into the surface of the soil before planting:

Diazinon	Root Guard
Phoxim	Fisons Soil Pest Killer
Pirimiphos-methyl	Sybol 2 dust

All Fruits

Nematodes or Eelworms
Dagger Nematode (*Xiphinema diversicaudatum*)
Leaf Eelworm (*Apphelenchoides fragariae and A. ritzemabosi*)
Needle Eelworm (*Longidorus elongatus*)
Stem Eelworm (*Ditylenchus dipsaci*)

The leaf eelworms feed in the buds and between the unfolded leaflets of strawberry. The fully expanded leaves are distorted and puckered, and rough grey or silver-coloured feeding areas are present near the veins. Fruiting is affected because the main crown is killed and replaced by a number of weak secondary crowns. The leaf eelworms are confined to strawberries and do not infest other crops.

The stem eelworm infests parsnip, rhubarb, beans, onions and strawberries. It persists in the soil and will attack these crops if they follow each other on the same ground. Symptoms on strawberry are thickening and shortening of the leaf and flower stalks; the leaf blades have a typical crumpled and ridged appearance and the plants are severely dwarfed.

The dagger and needle eelworms are soil inhabiting and feed on the roots of strawberries, other bush fruits and weeds. More serious, they transmit certain virus diseases from weeds to fruit crops and vice versa, and when a new fruit crop is planted in soil in which the preceding fruit crop was infected by virus disease, the new crop could become infected. When virus is not present, the physical feeding of large numbers of eelworms on the roots will reduce the vigour of the bushes.

Control Measures Plant certified stock to ensure that the plants are healthy in the first place. Uncertified stocks may contain virus and contaminate the soil.

Eelworms will only thrive and multiply if they have the roots of weeds to feed upon. Keeping fruit plots absolutely free from weeds reduces eelworm numbers and the risk of infection. Many of the eelworms, except stem eelworm, are unlikely to survive if the soil is kept absolutely free from weeds for four months between planting susceptible crops.

Removal of an infected crop after picking and sterilisation of the soil with Basamid* is the most effective treatment for eelworm. One ounce per sq yd (36 g/m^2) should be broadcast over and rotavated very thoroughly into the soil. The soil should then be rolled and if possible covered with a polythene sheet to prevent the fumes of the chemical escaping too quickly. Very precise instructions are issued with Basamid and these should be followed to the letter.

Raspberry Beetle (*Byturus tomentosus*)

Blackberry
Hybridberry
Raspberry

The raspberry beetle can be found infesting blackberries and hybrid berries as well as all raspberry varieties. The small brown beetles emerge from the soil during May and June. During the early part of the season they live in the tips of the new canes, eating the leaves, removing strips of tissue in a characteristic pattern. Later they feed on unopened flower buds and flowers and this gives rise to distorted berries. The beetles lay eggs in the flower and these hatch out into small white grubs that tunnel into and feed on the ripening berries.

Control Measures When large numbers of beetles can be found before flowering on the bushes, spray with insecticide and, to kill the white grubs, always spray as a matter of routine the first ripening berries as they turn pink, using:

	Available as
Fenitrothion	PBI Fenitrothion
or	
Malathion	Malathion Greenfly Killer
	or
or	Murphy Liquid Malathion
Rotenone	Liquid Derris

Raspberry Cane Midge (*Resseliella theobaldi*)

Raspberry

The death or dieback of fruiting canes in the spring is the most obvious symptom following infestation by cane midge during the

*Available in 11 lb (5 kg) packs from commercial sundriesmen.

previous summer. If the lower parts of the canes are carefully examined, a rough ridged cankered area will be found 2-4 in (5-10 cm) above soil level. Between 6-36 in (15-90 cm), dark coloured, slightly sunken irregular areas of dead tissue will be found. These areas are where, in the previous year, larvae of the midge fed under the bark of the new canes. The feeding areas become infected with various diseases that progressively kill the cane. (See Plate 21).

Control Measures To prevent this trouble, the new cane should be sprayed to kill the larvae of the midge before they damage the bark. Spray when splits in the bark of the new canes are seen, usually during the last week of May and again ten days later, making certain that the new canes are thoroughly covered with insecticide. Use:

	Available as
Fenitrothion	PBI Fenitrothion

Add 1⁄10 fl oz (3 cc) washing up liquid to each gallon of spray to get a better cover of the canes.

Vigour control, described on page 100, by which the first flush of new canes are removed, gives a very good control of this pest. The later and weaker growing canes do not split their bark and, therefore, do not provide egg laying sites for the adult midges.

Blackberry
Hybridberry
Raspberry

Raspberry Leaf and Bud Mite (*Phytopus gracilis*)

The raspberry leaf and bud mite is too small to be seen with the naked eye. It passes the winter under the bud scales and in April migrates to and feeds on the undersides of the newly opened leaves of the new and fruiting canes. A severe infestation gives rise to crumbly fruit, a reduced crop and stunted canes. The feeding on the leaves gives rise to distortion and irregular blotching of the leaves. To the inexperienced observer, these blotches are confused with virus infection. Outbreaks of this pest are worst where raspberries are grown in very sheltered situations near tall trees.

Control Measures No chemical is available to the gardener for the control of leaf and bud mite.

Always buy certified stocks to obviate the risk of importing the pest on planting stock.

Where bushes are severely attacked, the drastic measure of

cutting down to the ground all the canes and burning them, should be taken.

This means the complete loss of crop for one year. The only other alternative is to scrap the plot and replant with certified canes.

Raspberry Moth (*Lampronia rubiella*)

Raspberry

During March the caterpillars of the raspberry moth climb out of the soil on to the canes and eat their way into the buds and newly emerged laterals. If these buds are cut open, a small black-headed red caterpillar ⅓ in (8 mm) long or the tunnel it has bored, will be found. Each caterpillar may attack more than one bud.

By the end of April, the caterpillars leave the buds to pupate. The moth is flying during June, laying eggs in the open flowers. The caterpillars feed on the berries for a short time, after which they drop to the ground where they hibernate until the following spring.

Control Measures In the middle of March, soak the bases of the canes and the soil 12 in (30 cm) on both sides with tar oil, so that the caterpillars are killed as they emerge from the soil. Applications two or three years in succession usually eliminate this pest from a plot, provided there are no untreated raspberries nearby.

	Available as
Tar oil	ICI Clean Up or Murphy Mortegg

Red Spider Mite (*Tetranychus urticae*)

All Fruits

All the fruit crops can become seriously infested with red spider mite. The damage is worst and more prevalent on strawberry; it is less serious and frequent on gooseberry and currants. Its effect on cane fruits is minimal and then only in hot summers. (See Plate 22.)

The female mite survives the winter in the soil and under debris at the base of the bushes. It is bright red, small and, in spite of its colour, difficult to find. In spring the females move to the undersurfaces of the unfolding leaves, where they lay numbers of minute transluscent eggs. The eggs hatch out into mites which are a pale greenish-yellow colour with black markings. Under favourable weather conditions seven or eight generations can occur, so very large numbers of mites can soon be found feeding on the leaves.

The mites kill the leaves by sucking the sap from the cells. At the onset of an attack, these feeding marks can be distinguished on the upper surface of the leaves as a group of individual white spots. Later, in a severe attack the spots merge, the leaves lose their fresh green colour, turn brown and are useless to the plant. If an attack is not checked in time, there will be a serious decrease in the weight of fruit to be picked.

Control Measures A weekly watch should be kept for the initial signs of an attack, when a spray of malathion should be applied and repeated 14 days later, making sure to cover the undersurfaces of the leaves, using:

	Available as
Malathion	Malathion Greenfly Killer or
or	Murphy Liquid Malathion
Pirimiphos-methyl	Sybol 2

When an attack is severe, spray with dimethoate and repeat at 14 day intervals until all the mites have been killed.

Dimethoate	Boots Systemic Greenfly and Blackfly Killer or Murphy Systemic Insecticide

Raspberry
Strawberry

Rosy Rustic Moth (*Gortyna micacea*)

In May or June the canes of newly planted raspberry and maiden strawberry plants can suddenly wilt and die. One cause of these troubles can be found by slitting open the cane or cutting open the crown and inside will be found a dun-coloured caterpillar. This is the larvae of the rosy rustic moth which lays its eggs on the soil in early spring; they hatch out and bore into the nearest plant. Attacks are sporadic and are usually worse after hot dry summers. Nothing can be done to control this pest, so it is fortunate that the number of plants attacked is usually small.

Slugs and Snails

Blackcurrant
Blackberry
Hybrid Berry
Raspberry
Strawberry

There are too many species of slugs and snails attacking crops to make their description worthwhile. They cause most damage to strawberry fruits, eating the flesh and making them susceptible to grey mould which can then spread by contact to sound fruits. The smaller snail species climb up into blackcurrant bushes to feed on the leaves and berries. Occasionally slugs feed on the new canes of raspberries and hybrid berries, eating away large areas of bark. This weakens the canes so that they break easily or become infected by various fungi. Slugs also graze on newly planted strawberry runners checking their growth or killing them.

Control Measures Broadcast slug pellets over the strawberry bed at the beginning of June before strawing down. If, later in the month, glistening slug trails are seen on the soil or on the plants, make a second application at least seven days before the fruit is expected to ripen.

In gardens where slugs and snails are known to be numerous, usually on heavy calcareous soils, broadcast slug pellets under blackcurrant bushes at the grape stage of flower development. Preventive measures for cane fruits should only be taken when damage to the canes is seen to be occurring.

Apply Slug Guard at 1 oz per 60 sq yds (55g/100m^2) (30 pellets per sq yd (36/m^2).

Strawberry Blossom Weevil (*Anthonomus rubi*)

Strawberry

The strawberry blossom weevil – or elephant bug as it is sometimes called because of its long snout that looks like a trunk – appears to be present on many strawberries in the south of England but has a more localised distribution in the north. At flowering time, the presence of a number of capped flower buds which fail to open and shrivel, is the first sign of the presence of this pest. If these capped blossoms are carefully examined, small white grubs should be found inside the flower, feeding on the stamens and ovaries. The flower is prevented from opening by the adult weevil boring a hole in the stalk ⅓ in (1 cm) below the flower bud. (See Plate 23.)

Control Measures Thoroughly spray the flower trusses as soon as they emerge from the crown and repeat the spray 14 days later, using:

	Available as
Fenitrothion	PBI Fenitrothion
or	
Malathion	Malathion Greenfly Killer or Murphy Liquid Malathion

Strawberry

Strawberry Seed Beetle (*Harpalus rufipes et al.*)

The strawberry seed beetles are sometimes called black beetles or rain beetles. They are quite large, ½-¾ in (12-18 mm) in length, with black bodies and each species has legs of differing colours. During the daytime they hide under stones and leaves and come out to feed at night. The beetles remove the seeds from the fruit, damage and eat the flesh. Linnets also eat strawberry seeds but the birds usually remove the seeds without damaging the flesh and normally from the upper surface of the fruit. On the other hand, beetle damage is usually situated on the undersurface nearer the soil.

Control Measures Broadcast, before strawing, and repeat 14 days later:

	Available as
Methiocarb	Slug Guard

½ oz to 30 sq yds (15g/25m^2) (30 pellets per sq yd (36/m^2)).

Strawberry

Strawberry Rhynchites (*Caenorhinus germanicus*)

The rhynchites is a closely related insect of the elephant bug. It inserts its eggs into flower trusses or leaf stalks and then severs the stalks below the eggs, so that the truss or leaf wilts and eventually dies.

Control Measures Spray during the third week of May and repeat the application ten days later. It is too late to spray when the damage is seen, so that once this pest is known to be present, routine spraying should be carried out each year using:

	Available as
Fenitrothion	PBI Fenitrothion
or	
Malathion	Malathion Greenfly Killer or Murphy Liquid Malathion

Strawberry Tarsonemid Mite (*Tarsonemus fragariae*)

Strawberry

This is a microscopic insect that can just be seen with a 20 × lens and in recent years has become a quite common pest of strawberries. Signs of an attack can be seen in the spring but they are more severe in August after picking, particularly after a hot summer. The plants are stunted, the leaves small and wrinkled and turn brown.

Control Measures There is no insecticide available to the amateur gardener to control this pest.

The mite can be kept in check by burning the straw over the plants immediately after harvest. The straw should be picked up with a fork and spread over the plants and allowed to dry. On a day when there is a fair wind blowing down the rows, set fire to the straw on the windward side, so that there is a quick fire that burns the leaves without damaging the crowns.

If it is impossible to fire the bed, the only remedy is to dig up and replant the bed with a new certified stock.

Tortrix Moths (*Acleris comariana et al.*)

Blackcurrant
Blueberry
Raspberry
Strawberry

Caterpillars of tortrix moths are most damaging on strawberries, less damaging on blackcurrants, and the harm they cause cane fruits is negligible. On strawberries, the young caterpillars which have black heads and bodies that are white, green or grey, depending upon the species, cause most trouble during April and May. They spin the leaflets together with silken threads and feed on the leaf blades. A number of the caterpillars bore through the sepals and petals of the unopened flower buds and feed on ovaries, stigmas and stamens. The flowers open in the normal way but the berries are distorted and susceptible to botrytis grey mould infection.

On blackcurrants they live similarly on leaves at the tips of the new shoots, spinning the leaves together and killing the growing points as a result of which the shoots branch. This does not reduce the crop but it is a nuisance to have a lot of forked branches.

On cane fruits the caterpillars spin the leaves together at the tips of the new cane and the fruiting laterals. This is seldom harmful and shoots and laterals usually grow away from the trouble.

Control Measures It is important to spray strawberries early with insecticide before the caterpillars spin the leaves together and bore into the flower buds. Spray as the buds of the flower trusses separate from each other and repeat the application ten days later using:

	Available as
Malathion	Malathion Greenfly Killer
	or
or	Murphy Liquid Malathion
Fenitrothion	PBI Fenitrothion
or	
Permethrin	Bio Sprayday
	or
	Fisons Whitefly and Caterpillar Killer
	or
	Picket

Strawberry

Vine Weevil (*Otiorhynchus sulcatus*)

Vine weevils are ⅓ in (8 mm) long, black and covered with short yellow hairs. They feed on the margins of the leaves but this is rarely serious.

The adult weevil lays eggs from May onwards on the soil near strawberries growing in the open, in pots and in tubs. The eggs hatch out into white legless grubs, with brown heads. They feed on the strawberry roots, causing stunting and death of the plants.

Control Measures As soon as the grubs are found feeding on the roots, water pots or tubs with insecticide solution and apply a drench of 1 pint (500 cc) to each plant growing in the open, using:

	Available as
Pirimiphos-methyl	Sybol 2

Blackcurrant
Gooseberry
Redcurrant

Winter Moth (*Operophtera brumata*)

During the winter the adult moths lay eggs on blackcurrant, gooseberry and redcurrant bushes. These hatch out into green caterpillars that have paler lines along their sides. They bore into the developing buds and feed on the flowers and leaves until June. They are sometimes called looper caterpillars because of their characteristic method of walking.

Control Measures As soon as damage to the leaves is seen spray, using:

	Available as
Fenitrothion	PBI Fenitrothion
or	
Permethrin	Bio Sprayday or Fisons Whitefly and Caterpillar Killer or Picket

Wireworm (*Agriotes obscurus*)

Raspberry
Strawberry

Wireworms, the larvae of the click beetle, tunnel at ground level into new raspberry shoots and newly planted strawberry crowns and kill them. Although wireworms can be present in all soils, they are more likely to be found in gardens that, within the previous five years, were grass meadows before building took place.

Control Measures Before planting rake into the soil:

	Available as
Diazinon	Root Guard
or	
Phoxim	Fisons Soil Pest Killer
or	
Pirimiphos-methyl	Sybol 2 Dust

If plants are attacked after planting apply ¼ pint (125 cc) of dilute insecticide to the soil round each plant using:

Pirimiphos-methyl	Sybol 2 Liquid

Glossary and Abbreviations

Glossary

cat-nosed	distortion of berries – looking like a cat's face.
chimaera	plant composed of two genetically different types of tissue.
cordon	bush formed of one straight stem.
crown	a compressed stem.
double digging	digging the surface soil to a depth of 8-10 in (20-25 cm) and forking or spading 8-10 in (20-25 cm) below that.
double planted	two canes planted together in the same planting hole.
drupelet	individual segment of a raspberry, blackberry or Tayberry fruit.
fruiting spur	short lateral branch that bears a number of fruit buds or berries.
grown on	planted out in a sheltered nursery bed, usually for one year.
grubbing	digging up and destroying.
hardened off	acclimatised to outdoor weather conditions.
herbicide	a chemical used to suppress or kill weeds.
lateral	a shoot arising along a main branch or stem.
leading shoot	the shoot that extends a main branch.
leg	the main stem of a fruit bush.
maiden	strawberry or other plant in its first year of growth.
micro-propagation	the rapid multiplication of plants under sterile, completely artificial, laboratory conditions.

mist propagation — rooting cuttings in a humid atmosphere, usually provided automatically with water jets.

permeable — allowing water to pass through.

plug — the hard centre of a raspberry.

runner — a small strawberry plant.

scorching — burning or drying up of the leaves.

setting of the fruits — fertilisation of the ovaries that results in seed formation and swelling of the ovaries.

spur — short lateral branch that bears a number of fruit buds.

stolon — thin stem produced by strawberries, on the end of which plantlets form.

stool — a discrete clump of shoots, where all the shoots arise from one place.

stooled bush — a bush on which the branches all arise from one place.

strig — a stem that bears a number of red, white or black currant berries.

systemic — a term used to describe an insecticide or fungicide which on application will be translocated throughout the plant.

tipped back — shortened by pruning.

truss — a stem that bears a number of flowers or fruits.

virus — agent, causing systemic disease, too small to be seen except by means of electron microscopes, but transmissable by grafting, by animal vectors or by contact.

water shoot — soft vigorous stem, usually arising at ground level from a stem or roots.

wetting agent — soap-like chemical that enables a dilute spray to wet the waxy or hair surfaces of a plant (e.g. washing up liquid)

Abbreviations

EMRS — East Malling Research Station

LARS — Long Ashton Research Station

SCRI — Scottish Crop Research Institute

List of Horticultural Societies and Institutes

United Kingdom and Eire

National Council for Conservation of Plants and Gardens
Wisley Gardens
Ripley
Woking
Surrey

Royal Caledonian Society
1 West Newington Place
Edinburgh EH9 1QT

Royal Horticultural Society
80 Vincent Square
Westminster
London SW1P 2PE

Royal Horticultural Society Garden
Wisley
Woking
Surrey GU23 6QB

There are no specific soft fruit growers' associations in the UK and Eire, but the various Colleges of Agriculture and Horticulture can supply information to the amateur soft fruit grower.

Askham Bryan College of Agriculture and Horticulture
Askham Bryan
York YO2 3PR

Berkshire College of Agriculture
Hall Place
Burchetts Green
Maidenhead
Berks.

Cheshire College of Agriculture
Reaseheath
Nantwich CW5 6DF

East Suffolk College of Agriculture and Horticulture
Otley
Nr Ipswich
Suffolk

Greenmount Agricultural and Horticultural College
Antrim

Hadlow College of Agriculture and Horticulture
Hadlow
Tonbridge
Kent TN11 0AL

Hampshire College of Agriculture
Sparsholt
Winchester
Hants.

Herts College of Agriculture and Horticulture
Oaklands
St Albans
Herts. AL4 9JA

Isle of Ely College of Horticulture
Wisbech
Cambs. PE13 2JE

Kew School of Horticulture Royal Botanic Gardens
Kew
Richmond
Surrey TW9 3AB

Kildalton Agricultural and Horticultural College
Piltown
Co. Kilkenny
Eire

Lancashire College of Agriculture and Horticulture
Myerscough Hall
Bilsborrow
Preston
Lancs. PR3 0RY

Merrist Wood Agricultural College
Worplesdon
Nr Guildford
Surrey

National Botanic Gardens
Glasnevin
Dublin 9

Norfolk College of Agriculture and Horticulture
Department of Horticulture
North Burlingham
Norwich NR13 4TA

Pershore College of Horticulture
Pershore
Worcs.

Plumpton Agricultural College
Plumpton
Lewes
East Sussex BN7 3AE

Salesian College of Horticulture
Warrenstown
Drumree
Co. Meath

Somerset College of Agriculture and Horticulture
Cannington
Nr Bridgwater
Somerset

Warwickshire College of Agriculture
Moreton Hall
Moreton Morrell
Warwick CV35 9BL

Welsh College of Horticulture
Northop
Clywd
Wales

West Sussex College of Agriculture and Horticulture
Brinsbury
North Heath
Pulborough
West Sussex RH20 1DL

Writtle Agricultural College
Writtle
Chelmsford
Essex

USA

American Horticultural Society
Box 0105
Mount Vernon VA 22121

American Pomological Society
103 Tyson Building
University Park PA 16802

California Strawberry Advisory Board
Box 269
Watsonville CA 95076

Gardens for All
180 Flynn Avenue
Burlington VT 05401

Home Orchard Society
2511 S W Miles Street
Portland Oregon 97219

Kiwi Growers of California
Route 1 Box 445
Chico CA 95926

Massachusetts Horticultural Society
300 Massachusetts Avenue
Boston MA 02115

North American Blueberry Council
Box 166
Mavmora NJ 08223

North American Fruit Explorers
10 South 055 Madison Street
Hinsdale IL 60521

Pennsylvania Horticultural Society
325 Walnut St
Philadelphia PA 19106

Washington State Fruit Commission
1005 Tilton Drive
Yakima WA 98907

The Agricultural Extension Service provides information for the amateur soft fruit grower. Enquires to the following should be addressed to the Agricultural Extension Service.

New York State College of Agriculture and Life Sciences
Ithaca NY 14853

University of California
Davis CA 95616

University of Massachusetts
Amherst MA 01002

Rutgers University
New Brunswick NJ 08903

The Ohio State University
Columbus OH 43210

Michigan State University
East Lansing MI 48823

College of Agriculture
Pennsylvania State University
University Park PA 16802

University of Illinois
Urbana IL 61801

Irrigated Agricultural Research & Extension Center
Prosser WA 99350

USDA
Washington DC 20250
can supply publications on fruit growing and answer enquiries

Index